Environmental Transformations

From the depths of the oceans to the highest reaches of the atmosphere, the human impact on the environment is significant and undeniable. These forms of global and local environmental change collectively appear to signal the arrival of a new geological epoch: the Anthropocene. This is a geological era defined not by natural environmental fluctuations or meteorite impacts, but by the collective actions of humanity.

Environmental Transformations offers a concise and accessible introduction to the human practices and systems that sustain the Anthropocene. It combines accounts of the carbon cycle, global heat balances, entropy, hydrology, forest ecology and pedology, with theories of demography, war, industrial capitalism, urban development, state theory and behavioural psychology. This book charts the particular role of geography and geographers in studying environmental change and its human drivers. It provides a review of critical theories that can help to uncover the socio-economic and political factors that influence environmental change. It also explores key issues in contemporary environmental studies, such as resource use, water scarcity, climate change, industrial pollution and deforestation. These issues are 'mapped' through a series of geographical case studies to illustrate the particular value of geographical notions of space, place and scale, in uncovering the complex nature of environmental change in different socio-economic, political and cultural contexts. Finally, the book considers the different ways in which nations, communities and individuals around the world are adapting to environmental change in the twenty-first century.

Particular attention is given throughout to the uneven geographical opportunities that different communities have to adapt to environmental change and to the questions of social justice this situation raises. This book encourages students to engage in the scientific uncertainties that surround the study of environmental change, while also discussing both pessimistic and more optimistic views on the ability of humanity to address the environmental challenges of our current era.

Mark Whitehead is a professor of human geography at Aberystwyth University. His research interests include the politics of sustainable development, urban geography and environmental citizenship. He is the author of several books including *Spaces of Sustainability: geographical perspectives on the sustainable society* (Routledge, 2006) and *State Science and the Skies: governmentalities of the British atmosphere* (Wiley-Blackwell, 2009). He is the Managing Editor of the journal *Environmental Values*.

Environmental Transformations

A geography of the Anthropocene

Mark Whitehead

Routledge
Taylor & Francis Group

LONDON AND NEW YORK

First published 2014
by Routledge
2 Park Square, Milton Park, Abingdon, Oxon OX14 4RN

and by Routledge
711 Third Avenue, New York, NY 10017

Routledge is an imprint of the Taylor & Francis Group, an informa business

British Library Cataloguing in Publication Data
A catalogue record for this book is available from the British Library

Library of Congress Cataloging in Publication Data
  Whitehead, Mark
  Environmental transformations: a geography of the anthropocene/
  Mark Whitehead.
  pages cm
  Includes bibliographical references and index.
  1. Nature–Effect of human beings on. 2. Global environmental change. I. Title.
  GF75.W49 2014
  304.2–dc23
  2013035936

ISBN: 978-0-415-80983-2 (hbk)
ISBN: 978-0-415-80984-9 (pbk)
ISBN: 978-1-315-83267-8 (ebk)

Typeset in Minion and Univers
by Florence Production Ltd, Stoodleigh, Devon

For Sarah

Contents

Plates

Figures

Tables

Boxes

Acknowledgements

First, and foremost, I would like to acknowledge the support and insight of the numerous undergraduate and postgraduate students with whom I have shared the ideas contained within this volume over the last 12 years. Collectively they have helped me to refine my thinking on the changing nature of human relations with the environment and better understand how to teach about these issues. I am also in the intellectual debt of a number of colleagues who have patiently listened to my musings on the Anthropocene and its implications. I would like to thank Harriet Bulkeley, Sarah Davies, Carl Death, Deborah Dixon, Kevin Grove, Matt Hannah, Jesse Healey, Gareth Hoskins, Rachel Howell, Martin Jones, Rhys A. Jones, Rhys D. Jones, Rachel Lilley, Mark Macklin, Kelvin Mason, Peter Merriman, Simon Naylor, Jonathan Oldfield, Kim Peters, Jessica Pykett, Roger Keil and Mike Woods. I am also grateful to Esther and Rhodri at Y Ffarmers, for providing both delightful sustenance and a wonderful place to which I could retreat during the writing of this volume. Finally, I would like to acknowledge my family for all of their help and support. Particular thanks go to my daughters Anwen and Betsan; I am sorry that this book has kept me so busy, but together you bring great joy into my life and a renewed purpose to my writing.

CHAPTER ONE

Introduction

Geography in the Anthropocene

1

Welcome to the Anthropocene. It's a new geological era, so take a look around. A single species is in charge of the planet, altering its features almost at will. And what [is] more natural than to name this new era after the top of the range anthropoid, ourselves? (Pearce, 2007: 58).

1.1 MEME OR GEOLOGICAL EPOCH: INTRODUCING THE ANTHROPOCENE

It is, I must admit, unusual for an academic conference to produce a new geological epoch. But this is, in a sense, precisely what happened when the eminent Nobel Prize-winning atmospheric chemist Paul Crutzen announced the arrival of the Anthropocene. While attending a scientific conference, Paul Crutzen described suddenly feeling uneasy about a fellow delegate's use of the term Holocene (Pearce, 2007: 58). The Holocene is a geological term used by environmental scientists to denote the warmer, inter-glacial period in which we now live (the Holocene began approximately 12,000 years ago, or around 10000 BCE). For Crutzen, however, the rapid and extensive nature of global environmental change over more recent history made the Holocene seem like an out-dated marker. In a short article he would later write in the journal *Nature*, Crutzen explained why he felt we had experienced an

environmental shift of geological proportions (see Crutzen, 2002; see also Steffen et al, 2007).[1] At the heart of Crutzen's argument was his belief that humans had collectively become a force of nature. For Crutzen, what marked humans out as a force, at least, equivalent to nature were two key processes: 1) the range of different ways in which humans had transformed the environment; and 2) the ways in which these transformations were increasingly expressed at a planetary level. The Anthropocene is thus marked, according to Crutzen, by 'greenhouse gases' reaching their highest levels for 400,000 years; the increasing power of humans to regulate and control the flow of water through dam-building and sluice constructions; global industries releasing some 160 million tonnes of sulphur dioxide into the atmosphere each year; increasing levels of oceanic exploitation by the fisheries industry; rising rates of artificial fertilizer application to soils; and the increasingly high extraction of minerals and aggregates from the Earth's crust through mining.

This book is premised on the fact that the Anthropocene appears to require a change in the ways in which we study environmental transformations. Studying the *deep times* of other geological eras and epochs has required scientific skills that can capture and interpret the hidden records of environmental history (including fossils, rock samples and sediment cores). Studying the Anthropocene is, however, a real-time project

Box 1.1 The Anthropocene

The term Anthropocene is increasingly being used to describe the geological epoch in which we now live. The term is actually a neologism (or new word) made from a combination of the prefix *anthropo-* (of humankind) and the suffix *-cene* (from the ancient Greek for 'new': this term is regularly used to denote new geological era such as the Ceno*zoic*, Pleisto*cene* and the Holo*cene*).

It was the atmospheric chemist Paul Crutzen and his colleague Eugene Stoermer (an ecologist) who first coined the term Anthropocene. Although the word Anthropocene is relatively new, the idea of a geological era of humankind has actually been with us for some time. Back in 1873, for example, the Italian geologist Antonio Stoppani used the term 'anthropozoic era' to convey the increasing impact that human beings (and the broader processes of agricultural development and industrialization) were having on the global environment (see Crutzen, 2002).

Plate 1.1 Paul Crutzen
Source: Wikimedia Creative Commons, Biswarup Ganguly

Key readings

Biermann, F. et al (2012) 'Navigating the Anthropocene: Improving earth system governance', *Science* 335: 1306–1307

Crutzen, P.J. (2002) 'Geology of mankind', *Nature* 415.3: 23

Economist (2011) 'The Anthropocene: A man-made world', *The Economist* 26 May

Rohe, R.E. (1983) 'Man as geomorphic agent: Hydraulic mining in the American west', *Pacific Historian* 27: 5–16

Steffen, W. Crutzen, P.J. and McNeill, J.R. (2007) 'Are humans now overwhelming the great forces of nature?' *Ambio* 36(8): 614–621

Zalasiewicz, J. et al (2008) 'Are we now living in the Anthropocene?', *GSA Today* 18: 4–8

that requires us to look as much at the horizontal record of human–environmental relations across (and above) the surface of the planet (including habitat change, urban sprawl, coral bleaching and desertification), as at the vertical record of the geological past. But if we take Crutzen's notion seriously, studying the Anthropocene also requires much more of us. It means that we must have a reliable toolkit for studying the geological force that is humankind. Humans are very different objects of enquiry than the forces that have shaped and defined previous geological epochs. To understand them appears to require a peculiar mix of analytical skills spanning psychology, anthropology, economics, politics, history, sociology, biology and geography. Consequently, while understanding the nature of the environmental past has required an understanding of geological and paleontological processes (including extinction level events, the changing composition of the atmosphere and the movement of tectonic plates), studying the Anthropocene requires us to ask questions about the drivers of human behaviour, the structures of global capitalism, the processes of urbanization, the political constitution of nation states and the nature of multinational corporations.

Plate 1.2 A geological timeline of life on the planet Earth
Source: Getty Images

Scientists remain uncertain as to whether the human impact on the global environment constitutes a geological level shift in planetary history (see Zalasiewicz et al, 2008). One of the key issues is that for the 'age of humans' to exist geologically, it is necessary not only to show that humans have changed the environment (something that the fossil record for the Anthropocene, which will include things ranging from cities to an assemblage of domesticated animal life, should demonstrate (*Economist*, 2011)), but also to illustrate that humans actually changed the ways in which the global environment operated (this tends to be more difficult to discern from the relatively short-term perspective we currently have on the would-be Anthropocene). The International Commission on Stratigraphy (which essentially polices the official geological timeline of the Earth) has established an official Working Group to explore the scientific credentials of the Anthropocene (this group includes Paul Crutzen). It will be this Commission that will determine whether the Anthropocene is simply a popular meme, which has spread among academics and commentators as a helpful term, or a scientifically approved geological epoch (see *New York Times*, 2012).

While acknowledging these technical debates, this volume is primarily interested in what the processes of ecological change associated with the Anthropocene mean for those who study environmental issues. To these ends, whether

> Visit the website of the Anthropocene Working Group at: http://www.quaternary.stratigraphy.org.uk/workinggroups/anthropocene/.
>
> In addition to providing guidance on how the notion of the Anthropocene can pass from being an 'informal' to official geological era, the site also provides links to related articles about the Anthropocene.

the collective wisdom of scientists eventually determines that we are (or are not) living in new geological times is not the most important issue. The very fact that the International Commission on Stratigraphy is considering the Anthropocene's scientific validity suggests that something profound has happened in human–environmental relations. This volume provides an introductory account of the role of human beings, and associated social, economic and political processes, in transforming the environment. This book serves as an introduction on three counts: 1) it introduces the nature and extent of the physical changes human beings have caused to local and global environmental systems; 2) it introduces the different processes that appear to be driving environmental transformation; and 3) it asks what can be done, and what is being done, to address human impacts on the natural environment.

One aspect of the debates that surround the Anthropocene that is taken up by this volume are its ethical implications (for a broader discussion of the ethics of the Anthropocene, see Gibson-Graham and Roelvink, 2010). At one level the very idea of the Anthropocene presents an opportunity for humankind to collectively reflect upon its environmental impacts and responsibilities. To these ends, many in the environmental movement see the types of environmental transformation that are associated with the Anthropocene as a basis for reducing the demands we place on the planet, to challenge the assumed value of economic growth and to re-localize our economies. At the other end of the ethical spectrum are those who feel that the idea of humans as intelligent agents of geological power should be a spur to further and deeper interventions into our planetary ecosystem. For example, in its recent feature on the Anthropocene, *The Economist* (*Economist*, 2011: 6) observed that it was '[B]etter to embrace the Anthropocene's potential as a revolution in the way the Earth system works ... than to try to retreat onto a low-impact path that runs the risk of global immiseration'. *The Economist* review claimed that

Box 1.2 Geoengineering

Geoengineering is a term that is used to describe a series of large-scale, technologically driven interventions in the Earth's climatic system. Geoengineering can take many different forms, including: cloud whitening, space mirrors, carbon capture and storage. At the heart of all geoengineering efforts is a desire to artificially regulate the Earth's temperature and avert the onset of climate change. While supported by many as a necessary response to the emerging threats of climate change, many are sceptical of geoengineering efforts. Concerns have been expressed that a reliance on geoengineering could see less effort being made to reduce greenhouse gas emissions. Others, such as the World Economic Forum (2013), argue that geoengineering technologies carry with them the threat of being exploiting by rogue nations in order to cause climate-related problems in different parts of the world.

Key readings

See *The Guardian*'s special section on geoengineering at: http://www.guardian.co.uk/environment/
geoengineering
World Economics Forum (2013) *Global Risks 2013*, WEF, Davos

the same intellectual powers that had created the Anthropocene could enable the 'evolutionary leap' – particularly in the ways in which we harness energy sources and artificially regulate the global environment – that our contemporary environmental problems appear to require. At the heart of the technocentric solutions envisaged by *The Economist* were the building of zero-carbon energy infrastructures and the initiation of new geoengineering programmes. These *normative* perspectives on what we 'should' do in response to contemporary forms of environmental change, and where these perspectives come from, are themes we will continually revisit in this volume.

1.2 THE ROUGH GEOGRAPHIES OF THE ANTHROPOCENE

I find one of the debates surrounding the notion of the Anthropocene particularly interesting. This debate concerns precisely when this new geological epoch may have begun. Some see its origins in the human domestication of animals and the associated birth of modern agriculture (the

so-called *Long Anthropocene*) (see Chapters 3 and 7 of this volume). Paul Crutzen is more precise, suggesting that the Anthropocene began in 1784 when James Watt developed the first design for the steam engine and kick-started the industrial revolution. Others link the Anthropocene to the rise of nuclear technology and the clear radioactive traces it has left in the geological record. As a geographer, I feel that these historical deliberations can often lead us to forget an equally important question: where is the Anthropocene? Asking where is the Anthropocene is a spatial question. As a spatial question it has both historical and more contemporary implications. In historical terms, it leads us to ask in what places did the processes associated with the Anthropocene first begin? In more contemporary terms, it can result in important questions being asked about how the effects of the Anthropocene are being experienced differently in different locations. The geographical perspective I pursue in this volume means that the objectives set out above must become geographical questions. The key questions I thus explore are: 1) In what particular places have the changes that

Box 1.3 Reinventing Eden and the cultural roots of environmental values

In her book *Reinventing Eden: The Fate of Nature in Western Culture*, Carolyn Merchant provides a fascinating insight into the cultural roots of the optimistic and pessimistic human attitudes to environmental challenges (like those associated with the Anthropocene). In the context of Judeo-Christian discussions of the creation of the Garden of Eden, Merchant argues that the human expulsion from the Garden has at one level become synonymous with the start of a process of human development and betterment that is gradually transforming wilderness back into a productive garden (the so-called *Recovery Narrative*). Others see the human mastery of nature as a negative trajectory, resulting in the loss of original nature and ecological wisdom (the so-called *declensionist*, or slide-down, narrative). Crucially, while moving in opposite directions, Merchant sees these two narratives (one of enlightened human progress, the other of deep green environmental care) as encompassing stories of environmental recovery (either to a carefully managed garden or a return to a pristine wilderness). For Merchant it is important to consider stories that may lead in alternative directions and combine questions of social and ecological justice. On these terms, Merchant's work can be interpreted to suggest that in the Anthropocene there may be more future alternatives open to us than ones that point either to the creation of a world that is dominated by the needs of humans, or one that is only concerned with the welfare of nature (see Lorimer, 2012).

Plate 1.3 The Garden of Eden: The Expulsion from Paradise, nave mosaics from Palatine Chapel, Palermo, Sicily (mid-twelfth century)
Source: Wikimedia Creative Commons

Key readings

Lorimer, J. (2012) 'Multinatural geographies for the Anthropocene', *Progress in Human Geography* 36: 593–612
Merchant, C. (2004) *Reinventing Eden: The Fate of Nature in Western Culture*, Routledge, New York: 1–6
Thomas, K. (1984) *Man and Natural World: Changing Attitudes in England 1500–1800*, Penguin, London

human beings have caused to environmental systems been most acutely felt? 2) Where have the different processes that appear to be driving environmental transformation been orchestrated from? 3) How does what can be done, and what is being done, to address the human impacts on the natural environment vary from place to place?

Geographical perspectives on processes of environmental change are not only important because they supplement historical accounts of ecological transformation. They also have the potential to transform our understanding of the consequences of environmental change. In part, developing a geographical perspective on the Anthropocene means that we have to question some of the spatial assumptions associated with modern forms of environmentalism. For some time now, ecologists and environmentalists have been engaged in an important, and often hard-fought battle, to make us think of environmental systems in globally interconnected terms (see Botkin, 1992). Thinking about environmental issues in relation to interconnected global systems has, however, more than just scientific value. By invoking the notion of a *Spaceship Earth* the environmental movement has attempted to build a unified political project, based upon the common ecological fate we all share. There is, however, a problem with thinking about environmental problems in global terms alone. In his book *Planet Dialectics*, the German sociologist Wolfgang Sachs, argues that we need to be wary of global perspectives on environmental problems (Sachs, 1999). Sachs' suspicion of Spaceship Earth-type visions of the environment stems from his assertion that although we live in an interconnected biosphere, we experience very different ecological fates.

The unevenness of our environmental fates is expressed nowhere more clearly than in the case of climate change (see Sandberg and Sandberg, 2010). Studies now indicate that the inhabited areas that are most likely to suffer the worst impacts of climate change are Africa and South Asia. These are the geographical areas that will bear the brunt of flooding, loss of agricultural

productivity and the spread of climate-related diseases. These, of course, are also some of the places that are least responsible for the production of the climate change problem, and are least able to protect themselves from it its impacts (see Chapter 9). According to Collier (2010), these so-called *involuntary exposures* to climate change may not only have devastating environmental affects on vulnerable nations. The uneven impacts of climate change could also see the opening-up of new tracts of agricultural development in the global north, just as they are disappearing in the global south (Collier, 2010: 3–4). Such a process could make it increasingly difficult for what Collier has termed the 'Bottom Billion' of humankind (the 60 countries that have experienced no substantial growth in incomes over the last 30 years) to attain prosperity in the future. A geographical perspective on the Anthropocene is, in part at least, about exploring the ways in which globally significant forms of environmental change affect different places and different people in very different ways.

Environmental geography does, however, involve more than simply mapping out what De Blij has termed the 'rough geographies' of global environmental change. Environmental geography also involves the study of *spatial relations*, *spatial locations* and *spatial systems*. *Spatial relations* take two interconnected, but distinct, forms. First, there are the routes taken by trade, transport, communication and pollution, which form the geographical means in and through which our world is joined together. These types of spatialized relations can be expressed on a map as the lines between ports or the trajectories taken by air pollution. Second, and perhaps more importantly, there are the myriad political, economic, social, cultural and environmental processes that constitute the collection of relations in and through which specific spaces (such as cities, regions, neighbourhoods and even nation states) relate to other places. These types of spatial relations are more difficult to express visually as they include financial transactions,

Box 1.4 Spaceship Earth and why we are all astronauts

R. Buckminster Fuller was renowned for asking people if they would like to be an astronaut. His response to the would-be space traveller was to tell them that they were already astronauts. In 1968, R. Buckminster Fuller published the now famous book *Operating Manual for Spaceship Earth*. The notion of a spaceship-like Earth travelling alone through the voids of space had entered the popular imagination following the circulation of the first images of Earth taken from space in the 1960s. For Fuller, the metaphor of Spaceship Earth was a useful way of conveying both the finite resources that humanity had at its disposal, and the crucial role of environmental maintenance in ensuring the ongoing wellbeing of our shared home. According to Fuller, just as a space capsule had to be carefully monitored and regulated to ensure that it could support the lives of the astronauts who were on board, so too did our planetary system require management to ensure humanity's survival.

R. Buckminster Fuller saw his ideas celebrated at Disney's EPCOT (Experimental Prototype Community of Tomorrow) theme park. He was consulted on the design of its Spaceship Earth attraction (see below), which actually employs the principles he patented for the construction of geodesic domes. Fuller developed geodesic dome structures in order to replicate key natural design principles.

Plate 1.4 Spaceship Earth, EPCOT
Source: Wikimedia Creative Commons

For an engaging account of how human understandings of machines, such as spaceships, have influenced the ways in which we collectively think ecological systems work, and can be exploited, see Adam Curtis' three-part documentary *All Watched Over By Machines of Loving Grace*:

http://thoughtmaybe.com/browse/video/adam-curtis

forms of cultural exchange, and geopolitical struggle, which have occurred over long periods of time. According to Massey (2007: 13), a concern with this form of relational geography requires us to follow the '[l]ines of [a place's] engagement with elsewhere. Such lines of engagement are both part of what makes it what it is, and part of its effects'. In environmental terms, the relations that connect one place with another's environment are not only trans-boundary pollution (such as acid rain, toxic waste disposal and air pollution). Environmental relations are also hidden in the everyday products that we buy and consume. The prominent contemporary example of palm oil helps to illustrate this point.

Palm oil is an edible extract of palm trees, which is used in the production of a number of everyday products ranging from margarines to cosmetics. You may not recall seeing reference to palm oil on the labels of your supermarket purchases; this is because it is often classified under the more general term vegetable oil. Palm oil is also becoming an increasingly significant agro-fuel (Pichler, 2011). In 2010 and 2011 palm oil topped the global vegetable oils production list (with 53.3 million tonnes being produced during 2010–2011). One of the reasons that palm oil has become so popular in more economically developed countries is because it was believed that while it exhibited many of the same properties as hydrogenated fats (particularly in relation to its ability to harden and be used in the production of

baked products – like breakfast bars), it is was less likely to cause heart disease (see *Science Daily*, 2009). While the health benefits of palm oil are now being questioned, its role as a popular substitute for hydrogenated fats has already had serious environmental consequences. In countries such as Indonesia and Malaysia large swathes of rainforest have been cleared and peat wetlands drained in order to make way for profitable palm oil plantations (BBC, 2007). In addition to causing local environmental tensions (palm oil plantations are increasingly encroaching on orang-utan habitats), it has been argued by Greenpeace that the spread of the palm oil industry is contributing to the loss of crucial carbon sinks (in the form of both rainforests and peat wetlands), and thus contributing to enhanced forms of climate change. Ultimately, what the example of palm oil illustrates is that within the Anthropocene it is important to recognize the lines of geographical connection that join supermarkets in the UK with peat swamps in Indonesia, and the cooking and eating practises favoured in the US kitchen with global climatic change. Recognizing these often hidden forms of geographical association, and understanding why they connect the places they do in the ways they do, is a crucial step in beginning to critically interpret the nature of the Anthropocene (for an interesting discussion of relational space and emerging forms of relational empathy between different places see Rifkin, 2009).

But the study of spatial relations does not preclude the study of very specific spatial locations. Understanding the role of particular sites in the orchestration of the Anthropocene appears to me, at least, to be an important starting point from which to study how human–environment relations have developed in certain ways, and to think about how these relations could be transformed. Studying environmental locations in the Anthropocene is, however, about more than simply tracing our current environmental problems back to the early Egyptian farm (and the domestication of livestock), James Watt's workshop (and the invention of steam power) or the

---

Box 1.5 The North Pacific trash vortex as relational space

A troubling but nonetheless instructive place to think about the relational spaces of the Anthropocene is the North Pacific Ocean. Reports suggest that the ocean currents in the North Pacific (commonly referred to as the North Pacific Gyre) have produced two huge trash vortexes (the Eastern and Western Garbage Patches), made up of human waste products. Reports of the actual size of the trash vortex vary greatly, with estimates ranging from 700,000 to 15,000,000 square kilometres (see Marks, 2008). This floating mass of micro-plastics, cigarette lighters and syringes *inter alia* has been linked to the death of approximately 1 million seabirds every year and 100,000 marine mammals (Marks, 2008). The sheer scale of the North Pacific trash vortex provides a poignant place to think about human impacts on the marine environment within the Anthropocene. Given the fact that the trash vortex is formed by waste products that travel thousands of miles, from a multitude of places, before collectively forming this floating island of debris, the North Pacific also provides us with an indication of the complex relational processes in and through which environmental change is generated in the Anthropocene.

Key readings

Marks, K. (2008) 'The world's rubbish dump', *The Independent* 5 February, http://www.independent.co.uk/environment/green-living/the-worlds-rubbish-dump-a-tip-that-stretches-from-hawaii-to-japan-778016.html (accessed 1 July 2013)

Also watch the chilling documentary film *Midway: Message from the Gyre*, http://vimeo.com/25563376 (accessed 2 July 2013)

---

deserts of Los Alamos (the home of the Manhattan Project and the nuclear bomb). It is about understanding how certain places are both instrumental in orchestrating current human–environmental relations and at the same time express the effects of those same relations. Studying the ghost town of Pripyat in the Ukraine (and the effects of the Chernobyl nuclear disaster there), or Malakoff Diggins in California (and the impacts of large-scale hydraulic mining, see Chapter 2), or the Aral Sea in Kazakhstan and Uzbekistan (and its gradual retreat and desiccation), or the community of Bhopal, India (and the associated consequences of industrial catastrophes), provide important starting points for us to understand how collective forms of human–environmental relations are expressed and resisted in particular places. But locating the Anthropocene is not simply about exploring its most extreme geographical consequences. It should also involve

a concern for more mundane domestic places such as the home, the suburb and the slum (see Shove, 2003). Collectively, these places constitute critical contexts within which human relations with the environment get expressed and normalized. Ultimately locating the Anthropocene provides us with an opportunity to see if our theories about its nature and form are accurate, and the ways in which such general theories may apply differently in different places.

The final key component associated with geographical studies of human–environmental relations is its particular concern with the formation and operation of spatial systems. Spatial systems exist when geographical spaces of various sizes come together to form interconnected arrangements of coordination and support. Spatial systems can include cities, regional economies, nation states and even transnational economic and political blocs. In Harvey's classic work on urbanization, for example, he describes how the particular combination of housing, transport infrastructure, factories and offices mean that cities reflect a kind of spatial logic for capitalist society (Harvey, 1989a) (see Chapter 6). This form of integrated spatial logic can also be seen to operate at regional scales, where systems of knowledge and skills sharing facilitate the development of distinctive economic practices (see Scott, 1988). But spatial systems can also be see at the much more local neighbourhood levels, where home life, schools, working, recreation and shopping have to be organized and negotiated on a day-to-day basis. Perhaps the dominant spatial system in operation today is the global market place. This is essentially a system that is premised upon the assumption that the most effective, efficient and successful way of organizing economic activity is at a planetary scale (see Chapter 4). At this scale, it is claimed that investment can most easily support the most successful economic enterprises, and consumers can decide to purchase their goods from a wide range of transnational suppliers. Throughout this book we will consider how the spatial systems in and through which we organize our social, economic and political lives play a crucial role in shaping the kinds of relations that we have with local and global environments.

EXERCISE

Golden spikes and the Anthropocene

Plate 1.5 shows the *Ediacaran golden spike* in Australia. Golden spikes are used to provide visible markers between different geological periods. This particular spike is being used to define the Paleozoic era. For this exercise you must imagine that you are working for the International Palaeontological Congress. In this capacity you need to do three things: 1) draw up a list of three potential places where the golden spike marking the beginning of the Anthropocene could be placed; 2) provide a list of pros and cons for choosing each of the three sites you have identified; 3) nominate one of the sites as the preferred location for the golden spike and explain the reasoning behind this selection.

Plate 1.5 The Ediacaran golden spike, Australia
Source: Wikimedia Creative Commons

1.3 WHERE DO WE GO FROM HERE?

This volume is split into two main parts: *Environmental transformations* and *Living in the Anthropocene.* In the first part we explore the types of environmental change associated with the Anthropocene. The chapters in this part consider in turn questions of resources, soil, air, forests and cities. Each of these chapters follows a standard structure. The first part sets out the nature and extent of the environmental changes associated with these sectors. The second part sets out the key ideas and theories that can help

us to understand the processes that are behind contemporary patterns of environmental transformation. Respective chapters thus consider theories of resource use and availability, political ecology, science, globalization and urbanization. While these theories are particularly helpful for understanding the forms of environmental change that are discussed in the chapters in which they are introduced, they are ideas that can help us to interpret the Anthropocene more generally. The third part of these chapters focuses on particular geographical case studies of the themes in question. The geographical focus of these

Plate 1.6 Blog screen capture
Source: Mark Whitehead

chapters helps us to consider how the forms of environmental change associated in the Anthropocene affects different places in different ways and is caused by different sets of social, economic and political processes. The second part of this book – *Living in the Anthropocene* – considers the challenges and opportunities that exist to communities and individuals as they attempt to live with and adapt to environmental change. This part considers issues of environmental governance (and the role of the state in the Anthropocene), the nature of human environmental behaviours and how it may be possible to build societies that are better able to cope with environmental transformations.

Most books carry with them the very real danger that they will be out of date as soon as they are published. Given the rate of social, scientific and environmental change associated with the Anthropocene, this book is in more danger than most of being automatically out-dated. In order to try and militate against this danger I have developed a blog to complement this book. The *Placing the Anthropocene* blog reflects on contemporary developments that have relevance to the discussions covered in this volume. These themes range from discussions of the contemporary degrowth movement, to domestic energy use and the dangers of environmental league tables. The blog is supported by a Twitter feed that provides up-to-date links to relevant media stories. Below is an example of a blog entry that reflects upon the significance of explorations of Mars for our discussions of the Anthropocene.

FROM THE BLOG

'There were once two planets' – Martian chronicles for the Anthropocene

Posted on 12 June 2013

There were once two planets, new to the galaxy and inexperienced in life. Like fraternal twins they were born at the same time, about four and a half billion years ago, and took roughly the same shape … They were 'Goldilocks planets', our astronomers would say: just right for life (Bilger, 2013: 65).

Quite by chance, I was recently reading two reflections on our nearest planetary neighbour. On one day I commenced reading Ray Bradbury's melancholic, 1951 novel *The Martian Chronicles*. Set in a distant future, when humanity has successfully established colonies on Mars, Bradbury's novel reflects on the peculiar forms of human experience that emerge in this alien landscape. On the following day I read Burkhard Bilger's *Reporter at Large* piece for the *New Yorker*, entitled 'The Martian Chronicles: a new era of planetary exploration' (Bilger, 2013: 64–89). Bilger's piece offered an in-depth account of NASA's successful Curiosity Mission to Mars. His highly engaging narrative focused on two characters: Adam Stelzner (leader of Curiosity's entry, descent and landing team) and John Grotzinger (chief scientist for the Curiosity Mission). As Bilger pointed out, '[O]ne man wonders how to get to Mars, the other what we'll find there' (Bilger, 2013: 69). What I found most fascinating about this article was the backstory it provided on human exploration of the red planet. This is a history that encompasses Giovanni Schiaparelli's first astronomical mapping of Mars in the nineteenth century; the grainy images of the planet that were sent back from NASA's Mariner 4 probe in 1965; and the 40-odd spacecraft that have since been sent to Mars in the hope of unlocking its secrets.

What appears to connect Ray Bradbury and the Curiosity Mission is not a desire to better understand Mars per se, but to grasp more fully the nature of life on Earth. Just as science fiction reveals collective truths about the nature of human existence through the exploration of extreme socio-technical scenarios, the exploration of Mars appears to reflect a collective desire to comprehend Earthly ecologies. In this context, Mars appears to be an important, if perhaps unexpected, place in and through which to consider the nature of the Anthropocene. While the Earth and Mars appear so similar they have, of course, taken very different geological and ecological paths. As Bilger observers, 'By the time Earth took its first breath three billion years ago . . . Mars had been suffocating for a billion years' (Bilger, 2013: 66).

Literary reflections and scientific explorations of Mars are about opening up the dialectic of life and death that both connects and separates it from the Earth's own environmental history; it is about the (re)construction of a geo-ecological hypothetical; a 'what if things had been different?'! In relation to the Anthropocene the red planet thus looms menacingly: a salutary lesson in the contingency of the Earth's life giving potential. But the Curiosity Mission, with its elaborate Sky Crane and mind-blowing budget, is also suggestive of technological solutions to our current ecological predicament: the off-planet future that Bradbury was so keen to explore.

The news is now full of accounts of Curiosity's latest findings as it forages and drills on the Martian surface. As I read the updates on the mission I am reminded that finding life on Mars is a difficult task. It is for this reason that it is such an important place to think about the nature of the Anthropocene. Mars appears to be a place that is both furthest from and closest to our own ecological experience; as such, it may provide an extraterrestrial promontory from which to develop a new appreciation of life on Earth.

Key reading

Bilger, B. (2013) 'The Martian chronicles: A new era of planetary exploration', *The New Yorker* 22 April: 64–89

NOTE

1 Crutzen's first discussed the idea of the Anthropocene in an article that he co-authored with Eugene F. Stoermer: Crutzen, P.J. and Stoermer, E.F. (2000) 'The "Anthropocene"', *Global Change Newsletter* 41: 17–18. Although Crutzen popularized the notion of the Anthropocene, Eugene Stoermer was the first to actually coin the term.

KEY READINGS

Crutzen, P.J. (2002) 'Geology of mankind', *Nature* 415.3: 23. This short overview provides an accessible introduction to the idea of the Anthropocene and the different options it appears to present humanity with.

Lorimer, J. (2012) 'Multinatural geographies for the Anthropocene', *Progress in Human Geography* 36(5): 593–612. A more challenging, but important reflection on what the Anthropocene means for the ways in which geographers think about and study nature. Lorimer is keen to point out that the idea of the Anthropocene should not be used to simplify our contemporary environmental situation, and suggests that it could provide a framework within which to understand the multitude of different geographical ways in which the Anthropocene is emerging and being experienced throughout the world.

Steffen, W., Crutzen, P.J. and McNeill, J.R. (2007) 'Are humans now overwhelming the great forces of nature?', *Ambio* 36(8): 614–621. This paper provides a more detailed introduction to the forms of environmental change that are associated with the Anthropocene.

PART ONE

Environmental transformations

CHAPTER TWO

Resources

Oil and water

2.1 INTRODUCTION: THE SIMON–EHRLICH WAGER

The year is 1980 and two prominent academics are about to engage in a rather unusual exercise: a wager. One of these academics was the American biologist Paul Ehrlich. In 1968 Ehrlich had published an influential book entitled *The Population Bomb*, in which he predicted that rising levels of global population would lead to shortages in the availability of resources (such as food and energy) and significant forms of human suffering (Ehrlich, 1968). The other academic involved in this wager was Julian L. Simon. Simon was a professor of business administration, who was best known for his economic analysis of the relationship between population and resources. Simon's work contradicted that of Ehrlich to the extent that he claimed that population growth did not lead to resource scarcity. Moreover, Simon claimed that population growth provided a basis for renewed social innovation and for making more resources available (Simon, 1981). In order to resolve their academic dispute Simon proposed a wager to Ehrlich. The wager involved Ehrlich (actually he and his colleagues John Harte and John Holdren) betting on the price of five commodity metals (chromium, copper, nickel, tin and tungsten). Ehrlich's bet was that by 1990 the prices for these commodities would have increased (as you would expect if population growth was

increasing demand for resources), while Simon wagered that the prices for these metals would decrease in real terms (as you would expect if you believed that population growth would lead to social innovation and the rising availability of resources) (Ehrlich and Ehrlich, 1996). Simon ultimately won the wager of $1000: while the global population increased rapidly during the 1980s (by somewhere in the region of 800 million people), the adjusted prices of three of the five commodities went down over the same period (Ehrlich and Ehrlich, 1996). Ehrlich wrote Simon a cheque for $576.07 to settle the bet (a settlement relative to the prices that went up and down), which in some ways actually proved both parties to be right and wrong at the same time.

This story is significant because it provides important insights into the nature of one of the most pressing challenges and hotly debated issues associated with the Anthropocene: namely the long-term availability of environmental resources. One of the key ways in which the human influence over the environment can be seen in the Anthropocene is in relation to our collective use of the various resources that the planet provides us with. Environmental resources include the metals that were part of the Simon–Ehrlich wager, but also include other minerals and compounds that are mined and drilled for (such as coal, oil, natural gas and aggregates). These resources are used to provide the energy that is needed by society

Plate 2.1 Paul Ehrlich
Source: Wikimedia Creative Commons

to heat our homes and run our factories, but also to provide the materials we needs to build the products that we use on a daily basis (from homes to iPhones). But there are also other important resources upon which human society depends, such as fresh water and soil (see Chapter 3). A defining feature of the Anthropocene is humanity's emerging ability to extract, move and exploit resources at geologically significant levels (Crutzen, 2002). This situation has inevitably lead scientists such as Paul Ehrlich to question how long it may be possible to sustain humanity's increasing demands for environmental resources.

In crude terms, it is possible to characterize the positions of Paul Ehrlich and Julian Simon in relation to two broad schools of thought. Ehrlich is often associated with a *Malthusian* worldview, which suggests that in the contexts of rising population levels and the increasing relative demand for resources, that finite resources will eventually be exhausted (for more on Malthusian theories see section 2.3). Julian Simon by contrast is more *cornucopian* in his outlook. Cornucopians believe that technologies and market forces will work to ensure that in the future humans will have adequate resources to meet their collective needs (the phrase cornucopian derives from the Greek legend of the 'horn of plenty', which would supply its owner with all that they desired). This chapter explores these, and related, theories of resource use and supply. Ultimately it will be argued that neither provides an adequate account of the complex nature of human resource relations (Bridge and Wood, 2010). In this context, it is argued that while the cornucopians may be correct in recognizing the role of technology, politics and markets in regulating and redirecting the human use of resources, they fail to acknowledge the limits that do exist in the biosphere's ability to both continue to supply these resources and to absorb the pollution that often follows in the wake of resource use.[1] On the other side of the argument,

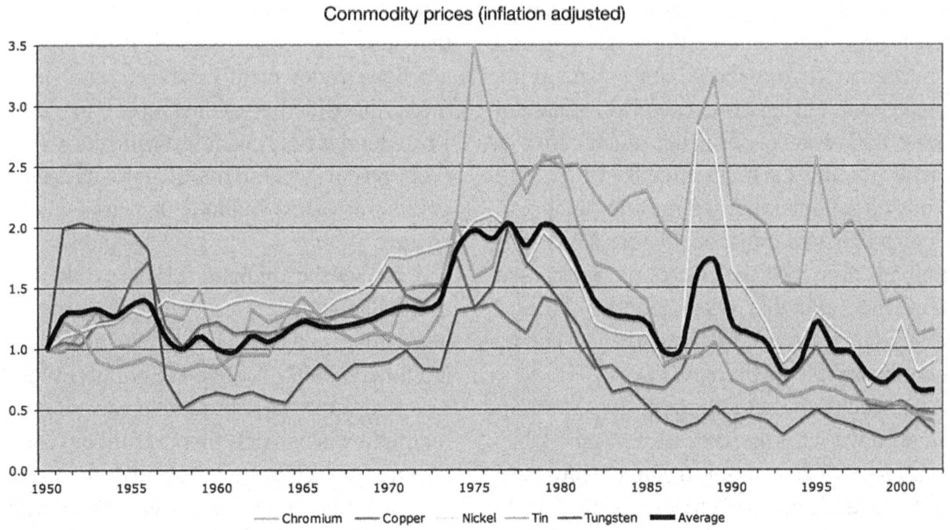

Commodity prices (inflation adjusted)

Figure 2.1 Fluctuating prices of the five commodities chosen in the Simon–Ehrlich Wager
Source: Wikimedia Creative Commons, author: Loren Cobb. Data source: US Geological Survey Data Series 140

we will see how Malthusian notions of absolute environmental limits can blind us to the complex processes by which resources are distributed and used.

This chapter begins with a brief analysis of current trends within human resource extraction and use. The chapter then moves on to explore theories of human–resource relations. Particular emphasis is placed in this section on the ways in which a geographical perspective can enable us to develop a better understanding of the nature of resource use and shortage. The final section of this chapter explores a specific geographical case study of environmental resource management: the issue of water supply in the Nile Basin.

2.2 CHANGING PATTERNS OF RESOURCE USE

In order to begin our analysis of resource use in the Anthropocene, this section sets out current patterns of human resource extraction and use. Analysis outlines the major resource groups that humans use, considers the geographical variations that exist in their relative levels of

extraction and utilization, and outlines some of the environmental consequences associated with the exploitation of key resources.

2.2.1 Mineral fuels

Mineral fuels are an area of resource use that has seen particularly pronounced increases during the Anthropocene. Mineral fuels include coal, peat, dry natural gas, natural gas liquids, petroleum and uranium. Mineral fuels such as these have provided the energy on which the Anthropocene has been based. These are the fuels that are used to power factories, heat homes, generate electricity and power the motor vehicles that we drive. Historically, coal and peat were the most significant fuels in the early stages of the Anthropocene. Coal has been used intermittently over significant stretches of human history, but it was not until the thirteenth century AD onwards we see it becoming a significant domestic energy source and basis for industrial energy production. Coal gradually replaced wood and charcoal as the main source of human energy and provided the basis for the industrial revolution in the eighteenth and nineteenth centuries in Europe and North America. From about the 1950s onwards oil and natural gas started to challenge coal's status as the dominant source of human energy. Oil and natural gas have the advantage of being much easier to extract from the Earth than coal (which still depends on labour intensive mining practices), and much easier to transport. Figure 2.2 reveals that oil has now replaced coal as the most-used mineral fuel, and that taken together oil and natural gas provide over half of all the fuel energy that is now utilized by humans.

What each of these mineral fuels has in common, however, is that they are non-renewable. By non-renewable, we simply mean that they have been formed over very long geological time periods, and once exhausted will not be available again for human use. In an influential book entitled *The Last Hours of Ancient Sunlight*, Tom Hartmann argued that our current use of energy-rich mineral fuels is akin to exhausting, within a couple of hundred years, energy resources that had been formed over millions of years of the Earth's history (Hartmann, 2001). As the energy that is locked up in mineral fuels such as oil, gas and coal

ultimately derives from the sun, Hartmann claims that our current fossil fuel economy is based upon a unique endowment of ancient sunlight. This means that with the exploitation of mineral energy, which has become possible during the Anthropocene, recent generations of humans have been granted access to millions of years of the sun's energy.

The fact that mineral fuels are only renewed over millions of years means that we are collectively likely to experience *peak production* moments (Heinberg, 2007). Unlike the notion of limits of resource supply, peak production does not refer to moments when society literally runs out of a given resource or commodity. Instead peak production relates to a point after which the rate of supply of a given resource gradually declines and the cost of the commodity increases (see Hemmingsen, 2010).

While we will discuss theories of peak production in section 2.3, for now it is important to note that the principles of peak production do not only apply to fossil fuels such as oil, natural gas and coal. Nuclear energy, which has been promoted by some as a long-term replacement of fossil fuels, is also likely to encounter a peak in its production

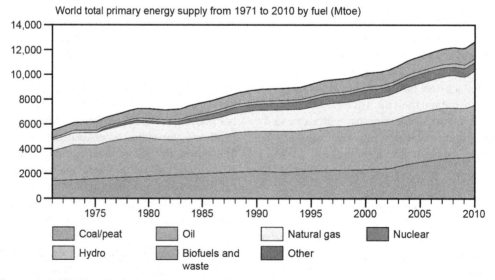

Figure 2.2 World total primary energy supply from 1971–2010 by fuel

Note: Mtoe: million tons of oil equivalent.

Source: International Energy Agency, 2012

potentials (Figure 2.3 indicates the increased use of nuclear power that has been evident from the 1970s onwards). While nuclear energy may well provide a more sustainable energy option than fossil fuels, the fact that nuclear power also relies on mineral fuels, such as uranium, means that it too will experience peak output limits (Heinberg, 2007).

Patterns of mineral fuel use and supply become more complex when they are considered in geographical terms. While the extraction, supply and use of mineral fuels is now a global affair, it is marked by significant geographical differences. Figure 2.4 reveals the regional shares of total energy use (in terms of both mineral fuels and all others) in 1973 and 2010. The figure illustrates

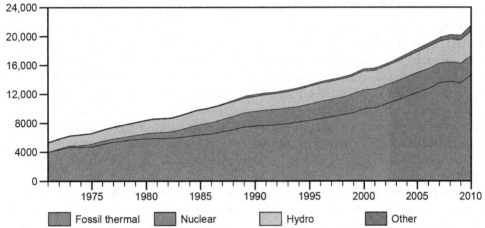

Figure 2.3 Fuel sources for world electricity generation

Note: TWh: terawatt hours.

Source: International Energy Agency, 2012

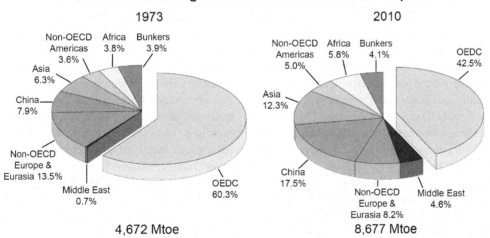

Figure 2.4 Regional shares of total global energy consumption: a comparison of 1973 and 2010

Source: International Energy Agency, 2012

two important things. First, that the affluent nations of the OECD (Organization for Economic Co-operation and Development), such as the UK, USA and Germany, have been the dominant consumes of global energy. Second, it illustrates that the OECD share of global energy use is starting to be reviled by rapidly expanding economies such as China and India (India is covered in Figure 2.4 under the category of Asia). This means that the world is currently in a situation that is characterized by dwindling non-renewable energy supplies and an increasing demand for such energy supplies from rapidly expanding economies.

The geography of contemporary mineral energy extraction and use is further complicated by the location of current oil reserves. While the assessment and predication of known oil reserves is a complex and often controversial exercise, we do have a clear sense of where the major reserves of oil are to be found. The largest known oil reserves are concentrated in the Middle East, in countries such as Saudi Arabia, Kuwait, Iraq, Iran and the United Arab Emirates. In fact, current estimates suggest that there are approximately 3.5 times more reserves of oil in the Middle East then there are in North America (the largest regional consumer of oil). Given the geographical concentration of oil reserves in the Middle East, it is unsurprising that the region has become so geopolitically significant in recent decades (Harvey, 2003).

While a more detailed discussion of the environmental problems that are associated with the use of mineral resources is provided later in this volume (see Chapter 3), it is important to briefly note some of these issues at this stage. The burning of mineral resources such as oil, gas and coal is the main human source of greenhouse gases at a global level. In addition to climate change, however, the burning of fossil fuels is also associated with a range of more localized air pollution issues. When fossil fuels are burned they not only release carbon dioxide but other pollutants such as ozone, nitrous

and sulphur oxides, and particulate matter. This cocktail of pollution is associated with the production of photochemical smogs and acid rain, which together contribute significant forms of ecological damage as well as having adverse affects on human health.

More recently, however, the extraction of so-called unconventional mineral energy resources has generated a series of new environmental problems. Unconventional mineral energy is a name that is given to a series of energy sources that are extracted using methods other than standard oil and gas well and rig technologies. Conventional methods of oil and gas extraction use wells and rigs to control the release of these resources at high pressures. These methods provide a relatively efficient and low-cost system of extraction. Unconventional energy sources include oil sands (which are a mixture of sand, clay water and petroleum), and shale oil and gas (in which oil and gas is distributed within the cracks and crevasses of sedimentary rocks). What unites these different unconventional energy sources is that the host materials that the oil and gas are found in (sand, clay and shale rock in these instances) prevent the fuels from flowing freely and thus being easily extracted. The removal of mineral energy from unconventional sources thus involves a series of processes that are both expensive to implement and often have harmful affects on the surrounding environment (Box 2.1). As the average price of mineral energy has increased over the last decade it has, however, become economically viable to extract unconventional sources of oil and gas. It is also important to note that as more and more unconventional mineral energy sources are being exploited (particularly in North America), the price of greenhouse gas-producing minerals such as coal is coming down! It appears likely that the exploitation of unconventional mineral resources is likely to contribute to the increased release of greenhouse gases from both unconventional and conventional sources.

Box 2.1 Fracking

Fracking (or hydraulic fracturing, to give it its full name) is a process that is deployed to extract shale gas from the ground. It involves a two-stage process. At the first stage engineers drill down into rock formations and lay a series of explosives. These explosives are used to breakdown (or 'fracture': from where fracking gets its name) the divisions that exist between different pockets of gas within the shale rock. The second stage of the process involves the injection of water, sand and chemical additives into the rock formations. This injection process means that pressure is applied to the gas deposits and that they can be released using well technologies.

The shale gas industry has been growing rapidly in the USA in recent years. The latest estimates suggest that the industry could expand from providing 15 per cent of total US gas supply (as it did in 2011) to 46 per cent in 2035 (EIA, 2011). Shale gas extraction is also starting to be developed in the UK in an attempt to secure new supplies of domestic energy. Shale gas extraction is controversial, however, because it is associated with a series of environmental side effects. There are concerns, for example, that potentially carcinogenic chemicals, which are used in the fracking process, could enter watercourses and pose a threat to human health and surrounding ecosystems. The use of explosives in the fracking process has also been linked with rising incidences of small earthquakes in the communities located near to fracking activities. Fracking activities carried out by the energy company Caudrilla in the UK have, for example, been linked to small earthquakes that were experienced in Lancashire in 2011.

Key reading

See BBC (2011) 'What is fracking and why is it controversial?', http://www.bbc.co.uk/news/uk-14432401 (accessed November 2012)

Gaslands

For a fascinating insight into the social and economic consequences of fracking watch Josh Fox's film *Gaslands*. In this documentary feature film, Josh Fox journeys across the USA to talk to people in the communities that have experienced first hand the side effects of this energy extraction process.

For an informative review of this film see: Bridge, G. (2012) 'Gaslands', *Area* 44: 388–390

2.2.2 Metal and non-metallic minerals

If energy resources are used to power our modern society it is metal and non-metal minerals that are used to build the things we use and consume in the modern world. Figure 2.5 illustrates contemporary levels of global production of non-metallic minerals. These non-metallic minerals include potash, phosphate and nitrogen in ammonia (which are used as fertilizers in agricultural systems – see Chapter 3), salt (which acts as a preservative in a range of modern foodstuffs and is used in a series of manufacturing processes) and sulphur (this is primarily used for the production of sulphuric acid, which is used in a range of commercial contexts including the production of agricultural fertilizers). The fact that society now requires 100 million tons of each of these minerals of an annual basis illustrates how resource dependent modern industrial society has become.

Figure 2.6 illustrates the production levels associated with a mineral that society is now producing at unprecedented levels: hydraulic cement. Since the mid-1980s the global production of hydraulic cement[2] has increased threefold to a contemporary level of approximately 3500 million tons a year (US Geological Survey, 2011). Cement is a binding agent that is widely used in the building industry (and in particular to produce concrete). Rising levels of cement usage reflect the significant building boom that modern economies have been going through over the last 40 years, as property development of various forms has become a cornerstone of urban economies (see Plate 2.2 and Chapter 6).

Figure 2.7 illustrates the changing global patterns in the production of selected metallic minerals. These key metallic resources have all shown aggregate increases in their levels of extraction since the mid-1980s, with iron ore (a key component in the iron and steel products that are widely used in construction and infrastructure projects) more than doubling in its levels of production between 1985 and 2010. The physical properties of metals mean that they can be used in a wide range of contexts. Metals with high levels of conductivity (such as copper) are used in the production of electrical products. The metal nickel is used in the production of stainless steel, rechargeable batteries, coins and even electric guitar strings. As humans continue to consume an ever-expanding range of consumer goods, from mobile phones and iPads to gold necklaces

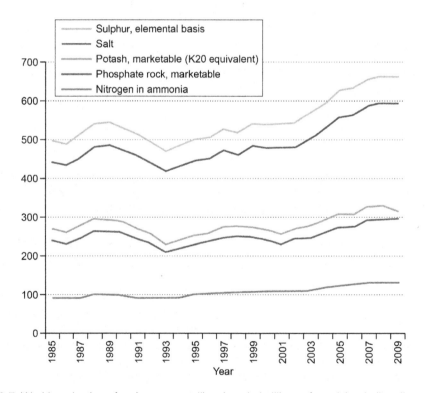

Figure 2.5 World production of major non-metallic minerals (millions of tons) (excluding diamonds and cement)

Source: Adapted from US Geological Survey, 2011

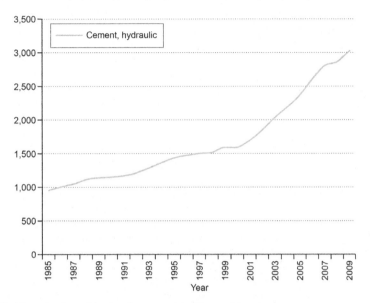

Figure 2.6 World production of hydraulic cement (millions of tons)
Source: US Geological Survey, 2011

Plate 2.2 The rise of the cement society: Spaghetti Junction, near Birmingham, UK

Note: Such complex road constructions would be difficult to achieve without the use of cement.

Source: Wikimedia Creative Commons, Highways Agency, 2009

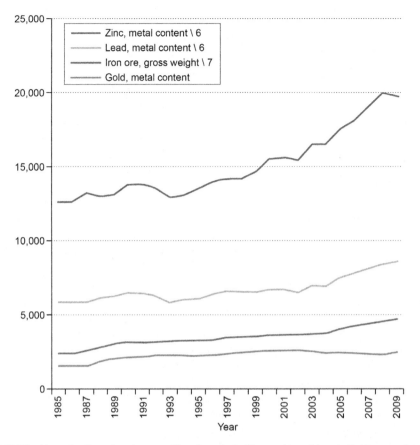

Figure 2.7 World production of major metallic minerals (millions of tons) (excluding cement)
Source: US Geological Survey, 2011

and Stratocaster guitars, the demand for metallic resources is likely to continue to increase.

What unites metallic and non-metallic resources is that they tend to be extracted from the Earth's crust using various mining techniques. These mining techniques are associated with a range of different environmental problems. At one level, surface level 'open cast' or 'strip mining' can totally destroy landscapes and the ecological systems that have developed upon them. At another level, however, the removal of minerals from the ground is often associated with the release of harmful pollutants into the environments that surround mines. The spoils (or leftovers) from mining activities often contain substances such as mercury and arsenic

(Plate 2.3). When rainwater passes over these mining spoils it can transport these pollutants into surrounding watercourses, which can have serious consequences for both human health and the sustainability of local ecological systems. As the demand for metallic and non-metallic resources continues to increase, the scale of associated mining activities is also expanding. It is in this context that we are now seeing the emergence of so-called super mines (or pits), which are able to extract resources from ground at greatly expended scales. The Anthropocene appears to be characterized by both increasing rates of mining and expansions in the size of individual mines. These two processes raise important issues concerning the long-term

Box 2.2 The rare earth dilemma

Rare earth metals are a group of metals that you have probably never heard of before; cerium and dysprosium are names that don't exactly slip off the tongue. Despite their obscure individual names, as a collection of resources they have featured prominently in political debates and media discussion over the last two years. As Figure 2.8 illustrates, between the summers of 2010 and 2011 the prices of rare earth metals increased rapidly. Given that rare earth metals are vital components within a range of modern products (including electronic goods and hybrid cars), it would be easy to think that the price increase associated with these resources was a product of increasing demand and declining availability. But the situation is more complex than this. Despite their name, rare earth metals are actually relatively abundant in the Earth's crust (*Washington Post*, 2012). The recent escalation in prices associated with these metals was thus not a product of scarcity, but of the fact that in recent years China has somewhat cornered the rare earth metals market (China produces somewhere in the region of 95 per cent of all rare earth metals on the global market place). This cornering of the market was in part achieved by China's ability to keep production costs down by only having limited 'environmental oversight' of its mining practices. The price hike in 2010/2011 was actually a product of the fact that China made the decision to restrict its export of rare earth metals (*Washington Post*, 2012). As there was only limited capacity to extract these metals elsewhere in the world, supply declined and prices inevitably went up. This price increase, however, suddenly meant that it became economically viable to restart rare earth metal production in other countries, which has led to increasing supply and declining prices for these metals since mid-2011 (see Figure 2.8). The example of rare earth metals reminds us that price alone is an unreliable indicator of absolute resource scarcity.

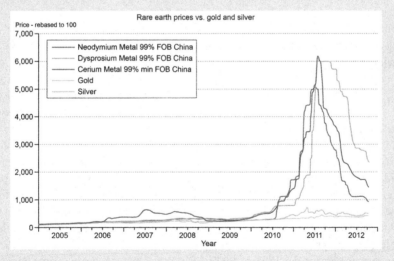

Figure 2.8 Rare earth metal prices versus gold and silver
Source: Reuters Graphic

Key reading

Washington Post (2012) 'China's grip on world rare earth market may be slipping', 19 October 2012

Plate 2.3 Mining spoil in the community of Blaenavon, South Wales
Source: Author's own collection

availability of key mineral resources and the growing environmental costs that are associated with their extraction.

2.2.3 Water

The final resource I wish to consider in this section is that of water. With the possible exception of air, water is the most important resource for our day-to-day survival. Water is used for drinking, sanitation, irrigation and various industrial processes. Although we only drink, on average, about 4 to 5 litres of water on a daily basis, global average levels of water consumption stand at about 150 litres per person per day (Pearce, 2007: 21). The additional 145 litres of non-drinking water we consume everyday are used to flush our toilets, fill our baths and operate our washing machines (among many other things). As with the energy and mineral resources we have mentioned so far

in this section, average levels of individual water use have been increasing during the Anthropocene. The major increases in rates of water use are associated with the need to produce more food to feed the expanding global population, but increases are also a product of new lifestyles, which have seen water being used to hydrate verdant suburban lawns and fill swimming pools (Pearce, 2007). But beyond these aggregate figures, there is a geography to increasing patterns of water consumption. In Australia, for example, the average per capita water use rates stand at 350 litres per day (that is 200 litres above the global average) (Pearce, 2007: 22). In the US, the situation is worse, with people consuming, on average, 400 litres of water, per capita, per day (Pearce, 2007).

At one level, it may seem that the rising levels of global water use should not be a concern in the same way that increasing levels of energy and mineral resources are. Water is, after all, a

Box 2.3 Malakoff Diggins and hydraulic gold mining

The case of Malakoff Diggins in California provides a troubling example of the environmental consequences that are often associated with mining. In the mid-nineteenth century, miners at Malakoff Diggins developed a new way of extracting gold from the land. Using a complex system of sluices, miners channelled water into powerful 'monitors' (or high-pressure hoses), which were used to blast rock and soil loose from the mine. Once these rocks and soil were freed, it became much easier for the miners to extract the precious gold that they were searching for. The new method of gold mining developed at Malakoff Diggins became known as hydraulic mining.

Plate 2.4 Water monitors used at Malakoff Diggins
Source: Gareth Hoskins

During the 1860s the environmental consequences associated with large-scale hydraulic mining became increasingly apparent. The water used in the mining processes at Malakoff Diggins repeatedly caused flooding downstream in the Sacramento River. These floods were in part caused by the 'tailings' (or spoil) that were carried in these mining waters, and resulted in the silting-up of the Sacramento River's bed. The downstream communities of Marysville and Yuba City were buried under 25 feet of mud as a direct consequence of hydraulic mining activities. Additionally, the floodwaters resulted in large swathes of agricultural land being covered in mine tailings, which led to the loss of valuable crops.

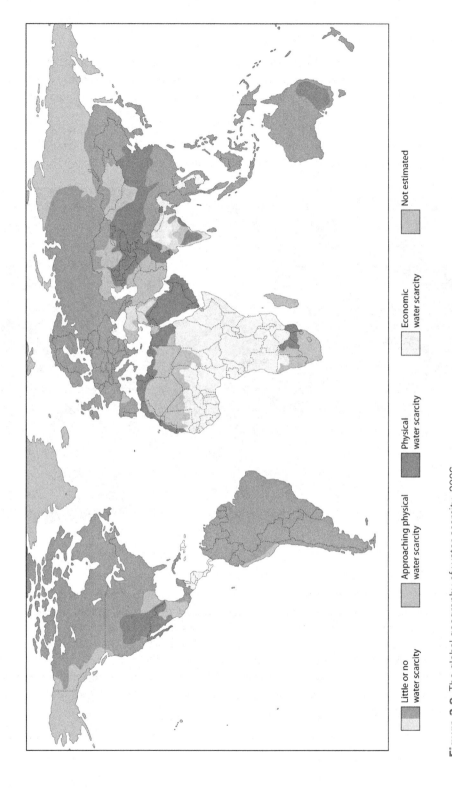

Figure 2.9 The global geography of water scarcity, 2006

Little or no
water scarcity

Approaching physical
water scarcity

Physical
water scarcity

Economic
water scarcity

Not estimated

Source: Wikimedia Creative Commons, after International Water Management Institute

renewable resource that is constantly been reused and recycled as part of the global water cycle. But the excessive use of water resources in particular regions is causing serious water shortage issues. This situation is further complicated by the potential impacts of climate change on water supply. Figure 2.9 illustrates key regions in the world that are already subject to water scarcity. Places such as the southwest US, North Africa, Southern Europe, Central Asia and the Middle East are currently most at risk from scarcity in the availability of fresh water. What is particularly troubling, however, is the fact that due to climate change these same regions are likely to see declines in water supply by somewhere in the region of 40 per cent. More worrying still is the fact that many of these areas are likely to see some of the largest increases in demand for water (see section 2.4 for a discussion of the situation in the Nile Basin). Taking these factors into account, it becomes apparent that water availability could be one of the most pressing resource issues of the twenty-first century.

2.3 DOOMSTERS, CORNUCOPIANS AND EVERYTHING IN BETWEEN

Having now established changing patterns in the human use of environmental resources, this section outlines some theories and ideas that can help us to understand these processes. While many of these theories directly contradict each other, collectively they can help us to understand the nature and likely consequences of the resource demands that we are now placing on the planet.

2.3.1 Parson Malthus and the neo-Malthusians

In 1798 a little-known English parson named Thomas Robert Malthus penned what would become one of the most influential statements on the limits that environmental resources could place on human development. Malthus famously described a key difference in the nature of population growth and the development of new resources. According to Malthus, while levels of resource discovery and availability tended to increase gradually, population tended to grow at much more rapid rates. According to Malthus, the rapid growth in population that he observed in eighteenth-century England would lead to an increasing demand for key resources (such as food, minerals and timber). Malthus was concerned that the ability to produce these resources could not match the demands of an expanding population. In this context, Malthus predicted that checks would be placed on human population growth. In Malthus's terms, these checks took two basic forms: *positive* and *negative*. Positive checks involved people proactively addressing the problems of excessive population growth through birth control and reducing the size of their families. Malthus's negative checks are more troubling. Negative checks are associated not with reduced birth rates but increased rates of death. In this context, Malthus argued that excessive population growth would ultimately lead to food shortages and hunger, and increases in human conflict and warfare as states scrambled to gain control of scarce resources.

Thomas Malthus's ideas have been widely criticized. At one level these criticisms have focused on the historical fact that Malthus's predictions have simply not come true: the global population has been able to rise to approximately 7 billion people (a figure that would have been inconceivable to Malthus), while the economy has grown to be 68 times bigger now than it was in 1800,[3] without being subject to the sustained forms of negative check. At another level, many have been critical of the underlying principles associated with Malthus's work (see Jackson, 2003; 2009: 6–13). Most crucially in this context, many have focused on Malthus's failure to take account of the role of technologies (from inorganic fertilizers to the use of modern mechanics) to increase the supply of resources.

Despite these criticisms, many modern scholars still broadly support Malthus's worldview (see Ehrlich, 1968; Homer-Dixon, 1999; Kaplan, 1994; Meadows et al, 1973; Ophuls, 1977; 1997; 2011). These so-called neo-Malthusians are united by a belief that despite the proven ability of technological developments to provide resources to an expanding global population, there are still real limits in the biosphere's ability to supply these resources and absorb the pollution that their use often entails. Donella and Dennis Meadows, working with a team of scientists at the Massachusetts Institute of Technology, produced one the most famous modern statements of neo-Malthusian work. The MIT team were able to bring Malthus into the computer age by using complex computational models to predict the likely resource implications of population growth. In their *Limits to Growth* report, Meadows et al predicated that economic development would be severely restricted if levels of population growth and resource use continued to expand (Meadows et al, 1973). Beyond computer models of future limits to growth, other neo-Malthusians have argued that the Malthusian future may already be with us. In his infamous account of demographic change and political struggle in West Africa, Robert Kaplan (1994) claimed that in certain parts of the world negative population checks were already evident. Kaplan argued that rising population levels and resource scarcity were leading to hunger and conflicts over scare resources in regions such as West Africa. Most troubling of all, Kaplan suggested that these positive checks on population growth were associated with the rise of anarchy in West Africa, and the associated breakdown of the rule of government and law.

For a detailed discussion of the *Limits to Growth* report and its political impacts go to:

http://www.clubofrome.org/?p=326

In his on going analysis of the connections that exist between politics and resource scarcity, William Ophuls (2011) suggests that what is happening in regions such as West Africa may provide salutary insights into what our collective political future may look like. According to Ophuls, over the last 200 years humanity has been experiencing an age of resource abundance. Ophuls claims that there is a connection between the unusually high levels of resource availability and the liberal, democratic systems that have emerged throughout large parts of the world (Ophuls, 2011). At the heart of liberal societies, such as those in North America and Europe, is the political principle that individual freedom and liberty are key features of a just and fair society (so long as these freedoms are not used to cause harm or reduce the freedom of others). Ophuls, however, recognizes that the formation of liberal societies has depended on resource abundance. Resource abundance has essentially enabled individuals and corporations to exploit resources without negating the ability of others to exploit available resources elsewhere. The free societies that many of us take for granted, are, according to Ophuls at least, dependent on there always being more resources elsewhere. As soon as resources become scarce (as Malthusians and neo-Malthusians argue they will), Ophuls claims that one person or corporation's exploitation of a resource will negate the free ability of others to exploit the very same resource (there will essentially be nowhere else to go!). In a situation of aggregate resource scarcity Ophuls claims that the world will be confronted with two possible futures. The first will see increasing conflict, and associated forms of anarchy, as people scrabble for available resources. The second will see the emergence of an increasingly authoritarian society, within which governments will have to restrict the economic freedoms of individuals in order to ensure that there are enough resources to go around. The neo-Malthusian future, it would appear, could be a rather undesirable mix of chaos and authoritarianism.

Box 2.4 Resource scarcity, conflict and 'the coming anarchy'

For neo-Malthusians such as Robert Kaplan there is a clear connection between the availability of resources (such as fuel and food) and human conflict. In his analysis of resource struggles in West Africa, he describes the ways in which population growth and the increasing scarcity of key resources has led to a rise in warfare and the unravelling of key social institutions (the so-called *scarce resource war hypothesis*) (Kaplan, 1994). In his recent analysis of so-called nature wars and resource conflicts, however, Le Billon claims that the relations between resource scarcity and conflict are complex and determined by important geographical factors (see Le Billon, 2001). According to Le Billon, the emergence of resource conflict, and the form that that conflict takes, is determined by the form and geographical location of the resource in question. One situation that does appear to result in the frequent emergence of resource conflict is when one state is overly dependent on a single resource (such as oil, diamonds or timber). In such instances it is relatively easy for a ruling elite to control that resource and become the primary beneficiaries of its exploitation (as we have recently see in relation to the Gaddafi regime's exploitation of Libyan oil). In such circumstances, it is common for rebel militias to engage in warfare in order to control the resource and either redistribute the wealth that is generated by its extraction or cement power and revenue for themselves.

The geographical location of a resource can also play a part in determining the types of conflict that may emerge around it. Resources tend to take two main forms: 1) *point resources*, which are concentrated in a small number of locations and are relatively easy to control by a ruling elite (such as gold mines and oil fields); and 2) *diffuse resources*, which are spread over wide areas and are difficult to effectively defend by state or military officials (such as agricultural land and forests). Le Billon argues that point resources tend to be associated with conflicts that attempt to overthrow the government, as gaining access to them means directly engaging the regimes that control them in something of an all-or-nothing conflict. Diffuse resources are, however, much more likely to be associated with warlordism (whereby military leaders establish their enclaves in an existing state), or, when the resource is located near the territorial edge of a state, with the formation of breakaway states. What Le Billon's work illustrates is that there is no simple connection between resources and conflict. First, it is clear that resource scarcity is not the only driver of conflict: the desire for power and wealth are also strong motivators. Second, it is also apparent that the type of conflict that emerges around resources is conditioned by the prevailing economic and political situation of the geographical region in question.

Key reading

Le Billon, P. (2001) 'The political ecology of war: Natural resources and armed conflicts', *Political Geography* 20: 561–584

2.3.2 Cornucopians and the Chicago Boys

At the other end of the spectrum to Malthus and neo-Malthusians are the so-called cornucopians. Unlike Malthusians, cornucopians do not believe that humanity is heading towards an environmental resource tragedy. While cornucopian thinking ranges very widely, it is characterized by two key features: 1) a belief that increases in population lead to the discovery and development of more resources; and 2) that the operations of the free market can help humanity avoid serious resource shortages. The first of these two assumptions is derived from the work of Ester Boserup, and is often referred to as *Boserupian*. The second assumption is most closely associated with a school of economic thinking that emerged from the work of a series of economists based at the University of Chicago in the post-World War II years, and is referred to as *neoliberalism* (see Peck, 2010). Let us now consider both of these perspectives and what they might mean for how we understand the relation between humans and environmental resources.

Ester Boserup was a Danish economist whose analysis of agricultural development is seen as one of the most significant challenges to Malthusian thinking. In her famous book *The Conditions of Agricultural Growth: The Economics of Agrarian Change under Population Pressure*, Boserup (1965) argued against the Malthusian understanding of the relation between population growth and food supply. Whereas Malthus claimed that population growth was controlled by food supply – with population growth being halted as soon as food production could not match increasing demand – Boserup argued that the supply of food was actually determined by population growth (Boserup, 1965). For Boserup, population played two important roles in historically observed patterns of increasing food production. First, increases in population created the incentive (namely more people demanding more food) for societies to develop the new practices and technologies that could increase food output. Second, the presence of more people, with their imaginative and creative potentials, increased the likelihood that new ways would be discovered to increase food production. These ideas have often been associated with Boserup's notion that 'necessity is the mother of invention'. It is important to note here, however, that there are a complex series of factors that are associated with the emergence of new technologies and inventions (including chance, opportunity, accident), of which necessity is only one factor (see Gladwell, 2009).

If Boserup's work can been seen as a celebration of human creativity, neoliberal cornucopians tend to emphasize the broader creative opportunities that are generated within free markets. Neoliberal thought is most closely associated with the influential work of Friedrich Hayek, Milton Friedman and their colleagues at the University of Chicago (Peck, 2010). The so-called 'Chicago Boys' (Klein, 2007) emphasize the great benefits that can be gained by organizing societies around free markets. In keeping with Boserup, the Chicago Boys claim that markets provide economic incentives for people to innovative – developing new products and technologies – as the competitive nature of markets mean that people are constantly looking for new forms of economic advantage. Neoliberals also recognize that through the changing prices of commodities (such as food, gold and oil), markets provide a very efficient way of telling producers which products are in greatest demand, and of providing extra investment (in the form of high prices) into resource sectors where there may be barriers to supply.

The work of Julian Simon (which we discussed in the introduction to this chapter) represents one of the clearest applications of neoliberal thinking to the analysis of resource use and supply. At the heart of Simon's argument is the assertion that the declining prices of many resources (such as the ones covered in his wager with Paul Ehrlich) illustrate the value of markets. According to Simon, it is free markets that both encourage innovation within, and ensure the supply of extra investment to, sectors where the supply of resources may be limited (and prices are high), and in the long term see these sectors produce more output (thus reducing the relative price of the commodity in question). In section 2.3.4 below, we will explore some significant critiques that have emerged to neoliberal analyses of human resource use.

2.3.3 Hubbert's curve and the peakists

We have already briefly discussed theories of peak resource production in this chapter. In this section

I want to consider related theories in greater detail and consider what they tell us about the changing nature of resource use.

Theories of peak resource production can be traced back to 1956 and to the work of a little-known geoscientist working for the Shell Corporation. In 1956, M. King Hubbert presented a paper within which he predicted that oil production in the US would reach a peak during the early 1970s. He would later go on to predict that the global production of oil would reach a peak in around 1995, after which point the supply of oil would enter a terminal pattern of long-term decline. While Hubbert's prediction concerning when the US would reach peak oil production proved to be surprisingly accurate, his global prediction has proved less reliable (there is no clear evidence that we have yet reached peak oil production). Notwithstanding this, many so-called peakists have argued that Hubbert's curve could be equally well applied to a range of other resources (see Heinberg, 2007).

M. King Hubbert explains his curve

This brief clip from 1976 shows Hubbert talking about his predictions for global peak energy supply. He discusses the potential impacts of oil cartels such as OPEC on oil production and argues that global peak oil production will occur at some time around 1995.

http://www.youtube.com/watch?v=lmV1voi41YY or search on Youtube for '1976 Hubbert peak'

At one level, it is tempting to interpret Hubbert's curve as just another, albeit more sophisticated, manifestation of Malthusian thinking: there are absolute limits in the availability and supply of non-renewable resources such as oil, coal and gas, and one day we will run out of these vital resources. In many ways, however, peak

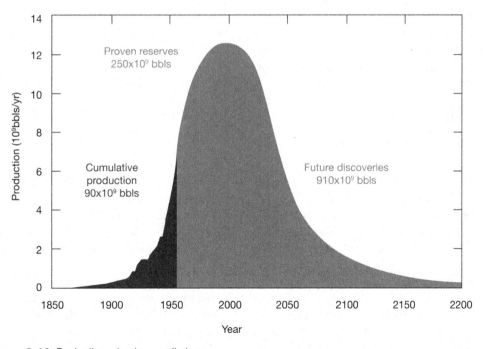

Figure 2.10 Peak oil production predictions
Source: Wikimedia Creative Commons: Hankwang at en.wikipedia

theories represent an interesting combination of geophysical accounts of absolute limits in resource reserves, and economic analyses of the moderating impacts of markets on resource supply and production (see Hemmingsen, 2010). On these terms it is significant that Hubbert's peak does not represent a prediction of when we will collectively run out of a resource, but is instead an account of when supply will reach a maximum. After that point, the market will be constrained in its ability to produce more of a given resource, but that does not mean that it cannot respond and adapt in other ways. As we have already mentioned, for example, as oil and gas prices rise (as we would expect them to do following the point of peak production), it is now becoming economically viable to explore the exploitation of unconventional energy sources through the controversial practices of fracking. While oil and gas prices remained relatively low, there remained little economic incentive to explore such expensive methods of energy extraction. In this context, Hubbert's curve could be seen to be as much a prediction of when new market forces will start to effect resource extraction and supply as a determination of when levels of peak production will occur (for an interesting account of the geographies of peak oil production see Bridge, 2010).

2.3.4 Marxism and the question of resource distribution

A final set of perspectives on human–resource relations is provided by a collection of work that is collectively referred to as Marxism. This work, as the name suggests, was inspired by the work of the nineteenth-century thinker Karl Marx (1818–1883) (for a colourful and engaging account of Marx's life and work see Wheen, 1999). Marxist studies of human–resource relations consider resource use in relation to broader systems of economic and political power. On these terms, Marxists are particularly concerned with the systems of ownership associated with resources; who benefits most from the exploitation of

resources; the labour relations that are associated with resource extraction; and the systems that distribute resources throughout society.

Marxist approaches to the study of human–resource relations are different to Malthusian accounts to the extent that they suggest that you cannot understand these relations simply by making reference to aggregate levels of population and resource availability (as Malthus does). For a Marxist, human–resource relations are instead the product of the complex processes that are associated with modern capitalism (see Harvey, 1977). It is in this context that Marxists are critical of both Malthusian and cornucopian perspectives on resource use and availability. Marxists claim that Malthusian concerns with resource shortages fail to recognize that there is often actually more than enough to go around. From a Marxist perspective, incidents of resource scarcity are often a product of the fact that the wealthy owners of a given resource tend to overexploit them for their own gain (perhaps by selling plentiful harvests to overseas markets, or placing oil revenues in overseas bank accounts). This means that local people see little of the resource that they may have helped to produce or extract. From this perspective, the forms of poverty and hunger predicted by Malthus are not the result of an *absolute* shortage of a resource, but a *relative* shortage in its availability to a local population. Furthermore, Marxists claim that Malthusian perspectives fail to account for the enduring forms of poverty that are a direct result of the unequal access and unjust distribution of resources among a population (Harvey, 1977: 232). By arguing that human poverty and suffering are the products of an inevitable shortage of available resources, Malthusians suggest that there may be very little that can be done to change this situation. Marxists, by contrast, claim that by changing the capitalist systems that determine who controls and profits most from resources, a miserable Malthusian future can be averted.

In some ways, Marxist perspectives echo those of cornucopians to the extent that they both claim that there are, potentially, enough resources to feed and sustain everyone. Marxists are, however, highly critical of the neoliberal assumptions that are often associated with cornucopian strategies. While cornucopians claim that the free market provides the ideal basis for enhancing resource production, Marxists claim that it is free market systems that are actually at the heart of contemporary social and environmental problems. In social terms, Marxists assert that it is free markets that have enabled the wealthy few to control vast swathes of the Earth's resources, and thus prevent their more even distribution. In environmental terms, so-called eco-Marxists claim that by giving prices to natural resources, free markets have resulted in the short-term overexploitation of resources in the pursuit of financial gain and a failure to appreciate the non-commercial value of the natural world (see Smith, 1984: Chapter 1).

2.4 WATER RESOURCES IN THE NILE BASIN

In the final section of this chapter we consider issues of environmental resource extraction, supply and use in the context of one particular resource in one particular geographical location. Geographically we focus on the Nile Basin in northeast Africa. In terms of environmental resources, we consider the challenges of water supply in this region. The Nile Basin is a large region encompassing 3350 million square kilometres of land and incorporating nine different countries (Burundi, Congo, Egypt, Ethiopia, Kenya, Rwanda, Sudan, Tanzania and Uganda) (Klare, 2002: 149) (see Figure 2.11). This relatively dry region is dependent on the water that is supplied through the Nile River system and its associated network of lakes and marshes.

During the colonial period of the nineteenth and early twentieth centuries, water use was tightly regulated by the imperial powers (in particular Britain) that controlled the Nile and its headwaters. Since the nations that are part of the Nile Basin have gradually gained independence, a series of processes have contrived to raise concerns over water scarcity in the region. At one level, the rising levels of population in the nations of the Nile Basin have placed strain on the ability of the river system to deliver adequate supplies of water. In 1950 there were approximately 60.5 million people living in the Nile Basin (see Klare, 2002: 157). By 1998, this figure had increased more than threefold to 206.6 million people (Klare, 2002: 157). But the story of population increases does not end there. Estimates from the World Resources Institute indicate that by 2050 the total population on the Nile Basin region will be 520.5 million (that is a 152 per cent increase on 1998 figures) (Klare, 2002: 157). Such rapid increases in population are naturally placing great pressure on water supply networks in the region. In particular, demand for water is rising as a consequence of the fact that many nations in the Nile Basin have, or are, developing new irrigation schemes. These irrigation schemes have been designed in order to enable desert areas to be transformed into agricultural zones that can provide food for the expanding population. Increasing demand for water is not only a product of increasing population, however, but also of relative increases in the level of per capita water use in the region. With many countries such as Egypt becoming increasingly urbanized, the amount of water that is being used per person is also increasing. This pattern of increasing water usage has been observed throughout the world as more economically developed, urbanized societies use water within modern sanitation networks, flush toilets and washing machines (see Kaika, 2005; Pearce, 2007).

Beyond increasing demand for water in the Nile Basin, other, geopolitical, issues have contributed to anxieties over future water scarcity in the region. These geopolitical tensions in the Nile Basin have resulted in different nation states in the basin developing new dams and irrigation schemes in order to enable them to increase their water withdrawal rates from the Nile (Klare, 2002: 157). As upstream states (such as Ethiopia and Sudan)

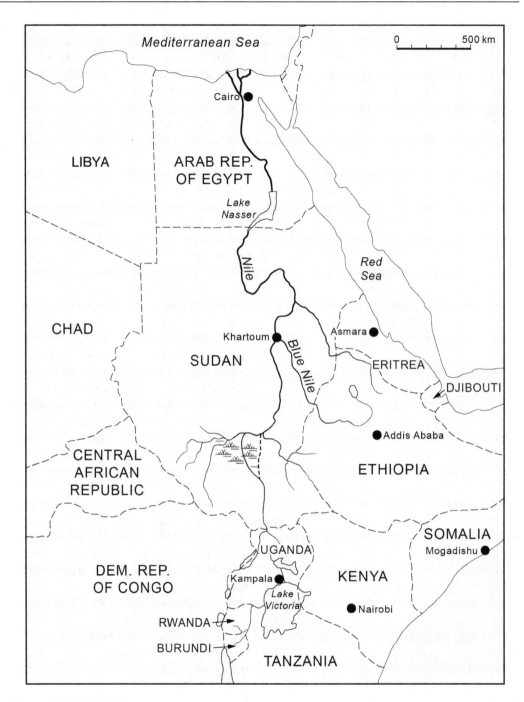

Figure 2.11 The Nile Basin
Source: Mark Whitehead

build more dams and irrigation schemes on the Nile, downstream states, such as Egypt, are becoming increasingly concerned about their own ability to be able to take more water from the Nile (Klare, 2002). Climate change could make the situation much worse. Recent estimates suggest that while water run-off may increase in upstream states in the Nile Basin (such as Sudan), it is likely to decrease in Egypt.

At present water withdrawals in Egypt and Sudan from the Nile and its tributaries is governed by the *1959 Nile Waters Agreement*. This agreement limits the amount of water that each state can withdraw from the river system in a year (with Sudan allowed to withdraw 18.5 billion cubic metres and Egypt 55.5 billion cubic metres

annually). In the recent past, when Sudan has suggested that it would withdraw more than its allocated 18.5 billion cubic metres of water from the Nile, Egypt threatened to use force to prevent such action. Given that Egypt is the dominant military force in the region, it continues to use the threat of military intervention as a way of

> The University of Michigan has produced a series of helpful web pages that explain the nature of water management issues in the Nile Basin:
>
> http://sitemaker.umich.edu/sec004_gp5/home

Plate 2.5 The Aswan High Dam (Egypt), a key part of Egypt's water security strategy
Source: Wikimedia Creative Commons, NASA

governing water use in the Nile Basin. In his book *Resource Wars*, however, Michael Klare (2002: 159) argues that as water scarcity becomes an issue of national security in the coming decades, the threat of military conflicts, or so-called water wars, could become a reality.

So what does the case of water use in the Nile Basin tell us about the nature of human–resource relations? At one level, the example of the Nile Basin appears to lend itself very easily to a Malthusian interpretation. As population growth outstrips the supply of resources such as food and water, society will be subjected to severe limits to its development and conflicts over such scarce resources may well ensue. The fact that population has continued to grow so rapidly in the Nile Basin, however, suggests that there might be merit in the cornucopian perspective. In this context it could be argued that population growth, and the pressure to withdraw more water from the Nile river system, has led to the development of technological solutions, including the construction of dams and innovative irrigation schemes that have enabled continued economic growth in the region. A Marxist perspective would, however, cast critical doubt on the nature of water scarcity in the Nile Basin and the ability of technological solutions to solve these problems. In the first instance, a Marxist analysis would explore the uneven distribution of water – both between states and different social classes – in the region. On these terms, Marxists may well point out that when there are droughts in the Nile Basin, it is the poor and not the wealthy that experience water scarcity and thirst. In the second instance, a Marxist perspective would question whether the building of dams and new irrigation systems in the Nile Basin is really solving water scarcity issues, and suggest that such initiatives may well be about securing water for the most powerful states (such as Egypt) and social groups (such as wealthy landowners).

However you wish to interpret water scarcity issues in the Nile Basin, it is clear that the region is an important place to consider the processes that are defining the Anthropocene. The Nile Basin

reflects two key features of the Anthropocene: 1) it is a place that is defined by the problems associated with the accelerated extraction of resources from the natural environment; and 2) it is a place where humans are trying to exert increasing technological control over the natural world (through the building of dams and irrigation systems). As we will see throughout the chapters of this volume, by considering the Anthropocene in places such as the Nile Basin it becomes possible to better understand the complex economic and political forces that shape human relations with the environment.

2.5 CONCLUSIONS

In this chapter we have considered changing patterns of resource use within the Anthropocene. Across a range of different sectors (including mineral fuels, metallic and non-metallic minerals, and water) we have observed patterns of accelerating use and consumption. While such patterns of resource use have not always led to scarcity, the extraction and use of these resource has had significant impacts on the environment: from the local problems of water pollution associated with gold mining, to the issues of global warming that have been caused by the burning of fossil fuels. Following our analysis of the changing patterns of resource use, this chapter has established a series of frameworks within which it is possible to interpret human–resource relations: Malthusian, cornucopian, peakist and Marxist. While contemporary policy-makers tend to favour market-based cornucopian understandings of human–resource relations, this chapter has shown that the use and relative scarcity of resources are complex issues that defy simplistic forms of analysis. A complex mix of availability, technology, political and economic power, market investments and exchange, and changing patters of consumption come together to determine resource use and scarcity. This chapter concluded by considering water scarcity and conflict in the Nile Basin. The case study helped to illustrate that when it comes to questions of human–resource relations it is often best to analyse them in the context of specific geographical locations and places.

NOTES

1 Related to this point Ehrlich did not feel that the five metals chosen as part of his wager with Simon were good indicators of the impending resource limits humanity was facing. He argued that fresh water availability, soil and forest resources were much more pertinent indicators of environmental limits (Ehrlich and Ehrlich, 1996).

2 Hydraulic cement hardens because of the chemical reactions that occur when it is added to water. Non-hydraulic cements must be kept dry in order to preserve their strength.

3 This figure is quoted by Jackson in his influential book *Prosperity Without Growth* (2009: 6). The figure was taken from Maddison's (2008) analysis of historical statistics on the global economy.

KEY READINGS

For a good overview of the key theories associated with the study of resource use see the opening chapters of Jackson, T. (2009) *Prosperity Without Growth: Economics for a Finite Planet*, Earthscan, London.

For an excellent, but slightly more advanced, analysis of resource scarcity issues see: Gavin Bridge and Andrew Wood (2010) 'Less is more: Specters of scarcity and the politics of resource access', *Geoforum* 41(4): 565–576.

For a detailed analysis of water supply and conflict in the Nile Basin see Klare, M. (2002) *Resource Wars: The New Landscape of Global Conflict*, Henry Holt Books, New York: 138–160.

Useful websites/blogs

To find statistics on energy use and availability (and the methods that are used to calculate them) go to the International Energy Authority: http://www.iea.org/stats/index.asp.

Chevron has produced an interactive online exercise through which you can explore the energy challenges that modern cities face: http://www.energyville.com/ (see how far you can get until Chevron tells you that your city needs oil).

CHAPTER THREE

Air

Science and the atmosphere

3.1 INTRODUCTION: THOMAS MIDGLEY AND THE ULTRAVIOLET CENTURY

When it comes to the large-scale environmental transformations that are associated with the Anthropocene it is, perhaps, unfair to single out individuals when apportioning responsibility for such changes. While it is evident that certain nations have disproportionately contributed to many of the environmental challenges we face today (see Sandberg and Sandberg, 2010), the idea that the actions of individuals can affect significant forms of ecological change appears farfetched. In relation to the atmosphere and the air that we breathe, however, one man has made an unusually significant contribution to its transformation. Thomas Midgley was an American scientist who worked at the interface of engineering and chemistry. Midgley was a highly respected and successful scientist in his day. He held over a hundred patents and helped to solve a series of problems that had baffled mechanical engineers for years. Despite these successes, in his *Something New Under the Sun*, John McNeill (2000: 111) claims that Midgley '[h]ad more impact on the atmosphere than any other single organism in Earth's history'.

Midgley's first major contribution to global atmospheric affairs came in 1921. While working for General Motors, Midgley discovered that adding lead to petrol could combat the problem of engine knocking[1] in motor vehicles. Midgley's discovery was so significant that between the 1920s and 1970s motor vehicles burned 25 trillion litres of leaded petrol (McNeill, 2000: 62). While the addition of lead to petrol was good for the operation of motor vehicles, it resulted in a significant spike in levels of lead in the atmosphere. This lead was able to enter human bloodstreams (where it is associated with metabolic disorders, hearing difficulties and the stunting of growth and development in young children) as well as ecological systems (where its absorption first into soils, and then plants, can see it entering the bloodstreams of various animals and causing a range of metabolic problems) (McNeill, 2000: 62). Following significant resistance from the car manufacturing and petroleum industries, it was not until the 1970s that significant steps were taken to regulate and eventually eradicate the presence of lead in petrol.

It was during the 1930s that Midgley laid the foundations for his second major contribution to human–atmospheric relations. Again at the request of General Motors (but this time their refrigeration division), Midgley turned his attention to improving the safety of fridges and cooling devices (McNeill, 2000: 112–113).[2] Midgley's major contribution in the field of refrigeration was to realize that the use of the chemical compounds of freon could improve the safety of domestic and industrial refrigeration. Freon is actually a brand name for a series of chlorofluorocarbons (hereafter CFCs), whose stable, non-flammable qualities made them ideally suited to the refrigeration industry. The problem is that CFCs are so stable that they only tend to break down in the upper reaches of the Earth's atmosphere, at which point they produce chemical reactions that erode the planet's protective layer of ozone. The Earth's ozone layer serves to protect the planet's surface from the most harmful forms of ultraviolet radiation that come from the sun. The

industrial use of CFCs only really began to accelerate after World War II, and by the 1970s it has been estimated that some 750,000 tons of the compounds were being emitted into the global atmosphere every year (McNeill, 2000: 113). These levels of CFC production led to two observed impacts on the planet's ozone layer: 1) a steady decline of concentrations of ozone throughout the world; and 2) a more marked seasonal decrease in ozone coverage in the polar regions (the so-called ozone hole) (Schiermeier, 2009) (see Plate 3.1). The loss of ozone in the Earth's atmosphere has

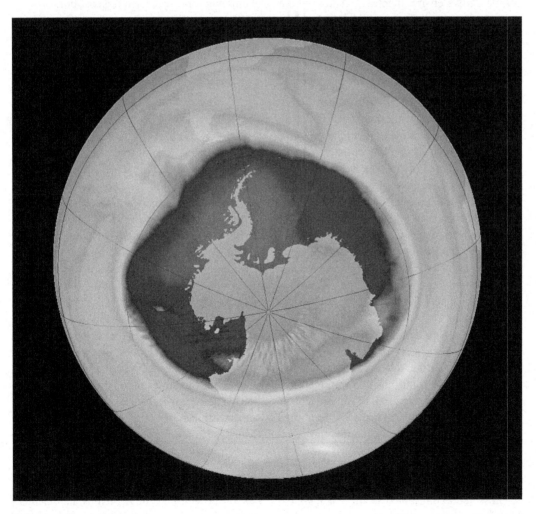

Plate 3.1 The ozone hole above Antarctica on 24 September 2006 was 11.4 million square miles
Source: Getty Images

been directly connected to elevated levels of skin cancer, terrestrial plant damage and dwindling levels of oceanic plankton populations.

Given that the observed socio-environmental effects of leaded petrol and CFCs only became apparent following Midgley's death in 1944, it is clearly unfair to apportion blame to the scientist. Notwithstanding this, the story of Midgley represents a cautionary tale of the, often, unintended consequences of scientific innovation and the care that must be taken to trial and monitor the impact on new industrial processes on the Earth's environment. In relation to this chapter, the story of Thomas Midgley provides a useful framework within which to position the discussions that follow. The atmospheric impacts of Midgley's infamous work can be broken down into two broad categories: 1) the *cumulative* forms of air pollution associated with leaded petrol; and 2) the systemic pollution associated with the erosion of the ozone layer. Cumulative forms of pollution relate to pollution that gradually accumulates in the environment over long periods of time, and contributes to various socio-ecological problems. *Systemic* forms of pollution, by contrast, refer to the ways in which certain pollutants can actually change the ways in which large-scale ecological systems (such as the ozone layer) operate. This chapter explores the complex mix of cumulative and systemic transformations of the atmosphere that have characterized the Anthropocene.

Midgley's story is, however, also important because it raises questions about the role of science and scientists in the Anthropocene. While Midgley's science contributed to severe atmospheric problems, the research of other scientists helped to identify these problems and prompted action to tackle them. Consequently, the international regulation of CFCs through the Montreal Protocol and the introduction of unleaded petrol were inspired by the studies of atmospheric scientists, and contributed to the gradual closing of the ozone hole and the phase out of lead pollution. It is in this context that this chapter considers the role of science and scientists in both generating atmospheric problems and in helping identify and address them (see Whitehead, 2009).

This chapter begins with an assessment of the atmospheric transformations that are associated with the Anthropocene. The second section outlines the nature of atmospheric science, and its role in mediating social relations with the atmosphere. The third section develops a geographical perspective on these issues by considering the role of scientists in the struggle for clean air in Louisiana's so-called chemical corridor.

3.2 A BRIEF HISTORY OF AIR POLLUTION: FROM MAUNA LOA TO MUMBAI

3.2.1 The Keeling Curve and the history of CO_2

Charles David Keeling was an American geochemist who was initially at the California Institute of Technology and the Scripps Institute of Oceanography. In many ways Keeling's scientific work represents the most significant contribution to human understandings of our collective impact on the global atmosphere. From early in his career Keeling had taken an interest in the seemingly innocuous gaseous compound carbon dioxide. CO_2 is technically a trace gas, which only comprises approximately 0.039445 per cent (or 394.35 parts per million by volume) of the Earth's atmosphere (although this level does vary on a seasonal basis). Despite the relatively small levels of carbon dioxide in the atmosphere, the gas has a peculiarly significant impact on life on Earth. The nineteenth-century scientists John Tyndall and Jean Baptiste Fourier had recognized that gases like carbon dioxide and methane trapped heat in the planet's atmosphere. It was not, however, until the meticulous nineteenth-century calculations of the Swedish chemist Svante Arrhenius that scientists began to realize the full environmental impacts of atmospheric CO_2 (Pearce, 2006: 33–35). In trying

Box 3.1 The greenhouse effect

The greenhouse effect is a naturally occurring process that is responsible for regulating the Earth's temperature. The greenhouse effect is a product of so-called greenhouse gases (such as carbon dioxide, methane, water vapour and ozone). It operates on the basis that the Earth's atmosphere allows solar radiation to pass through and to warm the planet's surface. The infrared radiation (heat) that is re-emitted by the Earth's surface is then trapped by greenhouses gases enabling the further heating of the Earth. In essence, greenhouse gases operate like a planetary blanket, which keeps the planet warm and suitable for life. Without greenhouse gases the Earth's average temperature would dip to a life constricting −18 degrees Celsius (IPCC, 2007: 97).

to solve the mystery of the onset and retreat of ice ages, Arrhenius calculated that fluctuations in gases such as carbon dioxide in the atmosphere could trigger significant enough shifts in global temperatures to account for geological forms of global warming and cooling. Arrhenius had discovered the greenhouse effect.

Before the work of Charles Keeling there was a popularly held belief that while greenhouse gases may regulate global temperatures there was very little humans could do to significantly disrupt this global balance and change the planet's climate. While working at Caltech, Keeling developed the first reliable instrument that could be used to accurately measure atmospheric carbon dioxide (Gillis, 2010). Keeling first used his instrument to measure atmospheric concentrations of carbon dioxide in California, but soon relocated his operation to the Mauna Loa volcano in Hawaii. Operating at some 3000 metres above sea level, Keeling used his Hawaiian base to assiduously monitor atmospheric carbon dioxide levels from 1958 onwards. What Keeling discovered would radically alter our understanding of what humans could actually do to global environmental systems, such as the greenhouse effect.

Keeling's painstaking work in the thin air of Mauna Loa revealed steadily increasing levels of carbon dioxide in the atmosphere. (When Keeling started his study he recorded carbon dioxide at levels of 310 parts per million, or to put things another way, every million pints of air in the atmosphere contained 310 pints of CO_2. As I write this chapter reports indicate that we have now reached the 400 parts per million threshold.) Collectively, his observations led to the production of the now-famous Keeling Curve (see Figure 3.1). This curve shows increasing levels of atmospheric concentrations of carbon dioxide over time, with smaller fluctuations of seasonal levels of CO_2.[3] While the National Science Foundation of the US terminated funding for Keeling's work in the early part of the 1960s (on the apparent basis that it was 'routine'), over time scientists and politicians became aware of its significance.

There are two implications of the Keeling Curve that have increased in significance during the final quarter of the twentieth century. Since the industrial revolution, scientists have been aware of rising levels of carbon dioxide (and other greenhouse gases) in the atmosphere (a consequence of the burning of fossil fuels). Recent estimates put the aggregate level of carbon that has entered the Earth's atmosphere since the industrial revolution at 200 billion tons (Pearce, 2007: 62) (that is approximately equivalent to the average weight of 41 billion African elephants!). Prevailing scientific theories of the time did, however, suggest that the Earth's oceans and ecosystems would eagerly absorb this additional carbon dioxide. The Keeling Curve clearly showed that not all of the additional CO_2 that was entering the atmosphere was being reabsorbed into the Earth's biosphere

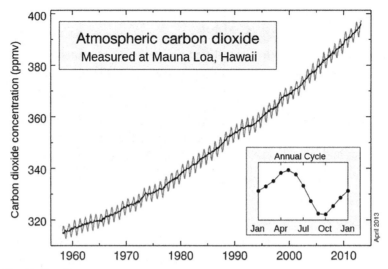

Figure 3.1 The Keeling Curve
Source: Wikimedia Creative Commons, Mauna Loa Observatory

and hydrosphere. The second implication of the Keeling Curve becomes apparent when it is correlated with measures of global average temperatures. The latest studies indicate that average global temperatures have increased by 1.5 degrees Celsius over the last 250 years (Rohde et al, 2012). As we explore below, while there are many possible reasons why global average temperatures may have increased so dramatically over the last 250 years (including solar activity), the work of Keeling (and the earlier work of Arrhenius) suggest that it is connected to rising levels of carbon dioxide in our atmosphere.

Later in this chapter we will discuss in more detail the relationship between carbon dioxide and climate change, and the controversies that still surround discussions of climate change. But before this, it is important to reflect upon other ways in

which humans have been reshaping the content of the atmosphere and the ways in which the global atmospheric system works.

3.2.2 Smoke, nuisance and the industrial atmosphere

When positioned in relation to the history of human–atmosphere relations, climate change is a relatively new manifestation of the problems associated with air pollution. The first clear signs of concern over human-induced air pollution can actually be traced back to fourteenth-century England (Brimblecombe, 1987). In 1306 King Edward I passed a Royal Proclamation banning the burning of sea coal in furnaces. This Proclamation was a response to the air pollution problems that were being generated in London by the growth of small-scale industrial premises in the metropolis (Whitehead, 2009). Unconfirmed reports suggest that in 1307 one resident of the city of London was executed as a consequence of disobeying the Proclamation. In many ways the Royal Proclamation of 1306 was a precursor to a long period of localized air pollution problems that were

> For an engaging discussion of the science and politics of climate change watch the former US Vice President Al Gore's film *An Inconvenient Truth*.

Box 3.2 Responding to climate change: mitigation and adaptation

As humanity has become increasingly aware of the threats posed by climate change, a series of responses have gradually emerged. The first set of responses is referred to as *mitigation* strategies. Mitigation strategies involve attempts to first stabilize and then reduce the human production of greenhouse gases. Mitigation strategies can take a variety of forms including the establishment of carbon taxes, international climate change agreements, carbon markets and trading schemes, carbon offsetting programmes and the development of low-carbon technologies. Attempts to develop climate change mitigation strategies have experienced a series of obstacles to their successful implementation. At an international level, it has proved difficult to establish international agreements on the how to collectively reduced global society's production of greenhouse gases. Although a United Nations Framework Convention on Climate Change has been established in order to coordinate action on climate change, individual states are concerned about the impacts that attempts to tackle global warming may have on their economic development prospects. At an individual level, it has also proved difficult to persuade people, who are used to the conveniences of personal car ownership and central heating, to adopt lower carbon lifestyles.

The second set of responses to climate change is collectively referred to as *adaptation* strategies. Adaptation strategies recognize that no matter how successful climate change mitigation strategies may be, society is now locked into certain forms of unavoidable climate change. Adaptation thus involves individuals, communities and states working to try to ensure that people are able to cope

Plate 3.2 The building of new flood defences on Borth Beach, Wales, UK
Source: Author's own collection

with the effects of climate change. Adaptation policies concern a range of policy areas including: flood protection, changes in agricultural production techniques, fresh water conservation, the construction of sea defences, responses to extreme weather events and the treatment of climate sensitive health issues, *inter alia* (see Plate 3.2). In 2010 the United Nations developed the Cancun Adaptation Framework to fund and support international climate adaptation measures. Increasing emphasis is now being placed on the development of so-called *enhanced adaptation*. Enhanced adaption focuses less on the implementation of adaptation schemes (like modern flood defences) and more on the development of local communities' ability to adapt to changing climatic circumstances. Such policies tend to focus on educational programmes and the promotion of indigenous practices.

While national government now tend to promote a mix of mitigation and adaptation policies, there are some tensions between the two policy areas. Those devoted primarily to mitigation efforts are concerned that too much investment in adaptation measures could result in less capacity existing to actually tackle climate change. Those who are now prioritizing adaptation methods argue that a narrow focus on mitigation policies could result in the already-existing plight of many climate-vulnerable people being ignored.

Key readings

Christoff, P. (2010) 'Cold climate at Copenhagen: China and the United States at COP15', *Environmental Politics* 19: 637–656

Sandberg, A. and Sandberg, T. (2010) (eds) *Climate Change – Who's Carrying the Burden? The Chilly Climates of the Global Environmental Dilemma*, Canadian Centre for Policy Alternatives, Ottawa

Whitehead, M. (2013) 'Neoliberal urban environmentalism and the adaptive city: Towards a critical urban theory and climate change', *Urban Studies* 50: 1348–1367

created by the concentration of ever-greater numbers of industrial premises within rapidly expanding cities throughout the world (see Chapter 6). On these terms, air pollution was associated with a range of artisanal activities that generated various public 'nuisances' in large medieval cities such as London and Paris. The smoke produced by the burning of coal and charcoal, the smells and odours emanating from slaughterhouses, and the acidic chemical compounds released from tanneries, collectively produced these atmospheric nuisances. But what made these air pollution activities problematic was that they were no longer occurring within isolated rural locations, but within bustling, and often overcrowded, urban communities where they posed a threat to human health and wellbeing.

In relation to London, it appears that the King's Royal Proclamation banning the burning of sea coal was, in the long term at least, unsuccessful. In 1661 the diarist John Evelyn published his famous observations on London's air pollution problems, *Fumifugium, or the Inconvenience of the Aer and Smoak of London Dissipated.*[4] Evelyn's pamphlet lamented the enduring air pollution problems of the city of London and the impacts that they were having on the general spirits of the population. The atmospheric problems documented by Evelyn would only get worse during the coming centuries with the onset of the industrial revolution in Britain. The industrial revolution of the late eighteenth and nineteenth centuries would have a profound impact on first Britain's and, ultimately, other countries' atmospheres. The industrial revolution ushered in an interconnected set of transformations in manufacturing, transportation and the practices of the domestic population. The rise of steam power, for example, revolutionized transport and manufacturing, as the burning of refined coal enabled the expansion of railways and the mechanization of the textile industries. The development of new techniques of iron and steel production in large-scale furnaces also required the burning of more and more coal. Finally, as Britain's population expanded, and more people moved into large cities in pursuit of industrial jobs, the burning of coal in domestic homes became much more prominent.

While the burning of more coal, to support steam powered equipment and to fire furnaces, generated black smoke, the industrial revolution also led to the production of new forms of atmospheric pollution. The emergence of alkali industries, which produced key chemicals that could be used in the manufacture of a range of substances including soap, glass and china, resulted in the elevated production of invisible air pollutants such as hydrogen chloride. When released into the atmosphere, hydrogen chloride forms hydrochloric acid, which can have a range of harmful effects on human health (including respiratory tract and pulmonary problems) and the environment.

3.2.3 Sulphuric atmospheres and transboundary pollution

The smoke and chlorides of the early industrial city were largely associated with local environmental problems. There was, however, an additional chemical in these polluting fogs that would contribute much more widely to the spread of ecological problems. Both coal and petroleum contain sulphur compounds. Consequently, when these fuels are burned they release sulphur dioxide (SO_2). For a long period of time the presence of SO_2 in the atmosphere was only believed to pose a threat to local air quality (Hajer, 1995: 126–127). As the burning of coal was gradually moved from the urban home to often rural power stations (with high smoke stacks), it was thought that the pressing threats associated with SO_2 production had been addressed. This false assumption meant that up until the 1960s people remained blissfully unaware of the ecological damage being produced by SO_2.

When SO_2 is present in the atmosphere it reacts with water molecules to produce sulphuric acid. Once produced, this highly corrosive acid can be transported over very long distances within weather systems before being deposited on terrestrial ecosystems by precipitation. This chemical process is now commonly referred to as acid rain (note that acid rain can also be produced

---

EXERCISE

Fumifugium

Download the full text of Evelyn's *Fumifugium* at:
http://ia600204.us.archive.org/6/items/fumifugium00eveluoft/fumifugium00eveluoft.pdf

When reading the text consider the following questions:

1. What motivated John Evelyn to produce this account of air pollution in seventeenth-century London?
2. What are the main problems that Evelyn identifies with air pollution?
3. What solutions does Evelyn recommend for solving London's air pollution problems?
4. Reflecting on the whole pamphlet, how does Evelyn's conception of the nature of air pollution differ from modern scientific understandings?

by nitrous oxides, which are also produced within the burning of coal and oil, reacting with water molecules to produce nitric acid). Various international scientific studies that were commissioned in the 1960s and 1970s began to show the significant environmental impacts associated with acid rain. Scientists in Scandinavia discovered that air pollution originating in Britain was travelling thousands of miles before being deposited as acid rain in Sweden and Norway (McNeill, 2000: 100). In the 1970s, studies conducted in the UK also illustrated that large areas of rural Britain were being affected by the acid rain originating from urban centres and power stations in the UK (Hajer, 1995: 128). North America has also suffered the consequences of acid rain. Here the prevailing transmission of acid rain was north from the US into Canada, with Canada receiving an estimated half of all its SO_2 pollution from the US (McNeill, 2000: 101). In Asia the geography of acid rain saw Japan being inundated with acid-bearing rain clouds from China and the Korean peninsula (McNeill, 2000).

Acid rain deposition is associated with a diverse range of ecological problems. When deposited in rivers and lakes, acid rain can severely erode an aquatic environment's levels of biodiversity. In this context, the acidification of water is toxic to fish, while the hatching of fish eggs is inhibited as water pH increases. In relation to forest systems, acid rain has the effect of weakening trees by attacking their leaves and inhibiting their ability to absorb nutrients (acid present in the soils of woodlands is know to dissolve nutrients stored there, meaning the minerals can be more easily washed away before being absorbed by tree roots). In addition to these ecological problems, acid rain is also responsible for the erosion of buildings and bridges and generates significant restoration costs.

3.2.4 Parasols of pollution: automobiles, smogs and particulates

In his book *Ecology of Fear*, Mike Davis (1999: 95) describes what has for many people become their most commonly encountered form of air pollution:

> Late August to early October is the infernal season in Los Angeles. Downtown is shrouded in acrid yellow smog while heat waves bellow down Wiltshire Boulevard. Outside air-conditioned skyscrapers, homeless people huddle miserably in every available shadow.

Two things are worthy of particular note in this excerpt. First, is the fact that Davis's observations come from Los Angeles. During the 1940s Los Angeles became the first home to a new form of atmospheric pollution: photochemical smog (see Plate 3.3). As a low-slung city of sprawl and suburb, Los Angeles was arguably the first city to be built for the automobile (see Chapter 6 for a more detailed discussion of this pattern of urbanization). Indeed, it was the combination of Los Angelinos' love of the motorcar and the geographical location of the city that generated this new form of air pollution. In order to understand the nature of this air pollution phenomenon, it is important to recognize a second feature of Davis's quote: its seasonality. While the infamous fogs of older industrial cities were traditionally associated with the cold air of winter, when high-pressure weather systems trapped air pollution that mixed with water vapour to produce thick fogs, by contrast, Los Angeles's fog is a phenomenon of the late summer.

Like their early industrial predecessors, photochemical smogs are produced under high-pressure weather conditions. Unlike older industrial fogs, where smoke pollution becomes trapped under a layer of warm air, photochemical smogs are a product of sunlight reacting with the nitrous oxides and volatile organic compounds (VOCs), which are emitted from vehicular exhausts, producing ground-level ozone and airborne particles. While harmful to human health in and of itself, this mix of ozone and airborne particles also provides a kind of catalytic chemical soup, which when it comes into contact with the sun's

Plate 3.3 Photochemical smog in Los Angeles
Source: Getty Images

rays can set off reactions that produce various additional VOCs and aldehydes. These chemically complex summer smogs are associated with asthma and various bronchial complaints. Cities such as Los Angeles, Mexico City, Teheran and Athens are particularly prone to incidents of photochemical smogs because they combine high car-use levels with warm climates and landscapes that actively contribute to the trapping of air pollution (McNeill, 2000: 72).

Box 3.3 Photochemical smog in Los Angeles

Los Angeles's location plays a crucial role in the development of photochemical smogs in the city. Its coastal location often means that onshore breezes return the city's air pollution to the metropolis as part of a cooler air mass (Plate 3.3). The cooler air mass pushes the warmer air it encounters in the city upwards. This process results in what meteorologists term an inversion layer. This inversion layer acts like a blanket trapping air pollution at low altitudes. Given that Los Angeles is surrounded by mountains, inversion layers can become very difficult to break when they settle over the urban region. These relatively stable atmospheric conditions provide time for the sunlight to stimulate chemical reactions at low altitudes within the city and the production of photochemical smogs.

Key reading

McNeill, J. (2000) *Something New Under the Sun: An Environmental History of the Twentieth Century*, Penguin, London: 72–76

A key component of smogs are airborne particulates. Particulate matter (PM) refers to any form of air pollution that is present in the atmosphere in either a solid or a liquid state. While the reaction of VOCs with sunlight produces some of the PM we associate with smogs, atmospheric particulates have a range of different sources. PM also includes the soot and black smoke that is produced from the burning of coal, the sulphuric acid suspensions that are associated with acid rain and various organic compounds. Particulate air pollution is often referred to as aerosols, although technically an aerosol is particulate matter and the gases in which it is suspended. Regardless of their precise name or form, PM has a series of common impacts. In terms of human health, atmospheric particulates have been associated with a range of conditions including heightened levels of asthma, acute respiratory infections, cardiopulmonary disease and cancers of the lungs and trachea (see Cohen et al, 2005). Such health problems are caused by the fact that a significant portion of the PM that is produced by air polluting activities is so small in size that it can penetrate the nose and throat of humans (where larger sized particles are normally caught) and enter throats and lungs. In a recent study for the World Health Organization, Cohen et al (2005) estimate that particulate air pollution is responsible for approximately 0.8 million premature deaths and 6.4 million years of life lost, per year.

Beyond human health, PM also has a significant impact on local climates. The particulates in smokes and smogs can act as a 'parasol of pollution', which prevents the sun's rays penetrating cities (see Pearce, 2007: 178). This parasol effect was associated in earlier industrial times with the rise of rickets within urban populations, which was believed to be caused by a deficiency of vitamin C in the population (vitamin C is naturally absorbed into the human body through the skin's exposure to sunlight). In today's world, however, the parasol of pollution effect is associated with the artificial cooling of cities that are thus protected from some of the worst consequences of global warming. In this sense, the particulate pollution of a city acts as a kind of greenhouse gas effect in reverse, preventing the sun's rays reaching the Earth's surface and being trapped in the greenhouse. This peculiar situation does, however, create something of a policy paradox. As urban authorities attempt to reduce the aerosols that are present in the atmosphere of cities, they may be releasing the full effects of climate change (it is estimated that aerosol pollution may be responsible for holding back around 0.2 degrees Celsius of warming, a quarter of the present recorded warming) (see Pearce, 2007: 179). The case of particulate air pollution reveals the complex relationships that exist between climate change and more localized forms of atmospheric pollution. It also reveals the ways in which localized air pollution is connected to systemic shifts in the ways in which the Earth's climate system works. To these ends, the production of ground-level ozone causes smogs and human health problems, whereas ozone in the upper atmosphere protects the Earth from dangerous forms of solar radiation. At a global level, the burning of fossil fuels produces an enhanced greenhouse effect that is contributing towards dangerous forms of climate change. At a more local level, the burning of the very same fossil fuels is generating palls of pollution that threaten human health, but shelter cities from the worst affects of climate change. It is thus important to remember that while the atmospheric changes that are associated with the Anthropocene may take local and global forms, they are all ultimately part of an integrated atmospheric system (for an interesting discussion of the complex relations that connect climate change and ozone depletion see Schiermeier, 2009).

3.3 REFLECTIONS ON THE NATURE OF ATMOSPHERIC SCIENCE

At the beginning of this chapter we saw the dual role that science plays in contributing to the atmospheric transformations that are associated with the Anthropocene and in alerting society to the dangers of air pollution and climate change. This section considers in more detail the nature of science and its particular role in mediating the relationships between humans and the atmosphere within which we collectively live.

3.3.1 On the nature of modern science

Of all of the Earth's spheres, the atmosphere is the most transitory. A complex mix of gases and water vapour, it is in a constant state of flux. This flux is driven at the local level by pressure systems and the weather. At larger scales, this flux can be seen within the seasonal transfers of carbon between the atmosphere and the biosphere. Taken together, the ever-changing nature of the atmosphere makes it very difficult for humans to know with certainty the precise nature of the fluctuations that are associated with this complex system. It was in this confusing atmospheric context that the certitudes of science appeared to offer humanity an effective basis for better understanding its relationship with the air.

The onset of the scientific revolution in sixteenth-century Europe resulted in the emergence of methods of observation and measurement that promised a new basis for humans to perceive and study the environments that surrounded them. At its heart, the scientific revolution rejected the power of religion and monarchical rulers as sources of absolute knowledge, and suggested that true wisdom could only be obtained through the dispassionate and careful scientific observation of the world (see Merchant, 1990; Shapin and Schaffer, 1985). Modern science has two key characteristics: 1) a commitment to objectivity; and 2) a quest for a universal basis for the study of nature (Haraway, 1991: 183–201). The objectivity, or neutrality, of the scientist is seen as crucial to ensuring that the findings of science are not corrupted by personal bias or political manipulations. The idea of universal, or standardized,

methods is an important part of the modern scientific tradition because it enables the experimental findings of one scientist to be compared (and then verified or nullified) with the work of other scientists (Hacking, 1983: 1–17).

Taken together, the objectivity and universality of modern science are important because they enable the production of more reliable and trustworthy accounts of the real world. It is crucial at this point, however, not to conflate the reliability of scientific studies of the real world with the notion of truth. At one level, scientists regularly make mistakes and misunderstand the things they observe (scientists originally thought, for example, that the worst heath effects of air pollution could be prevented if humans breathed through their noses rather than their mouths. The fine-grained nature of many particulates, however, meant that this precaution made little difference to the impacts of air pollution on human health). At another, deeper level, however, the nature of modern science means that it is constantly seeking to challenge accepted truths in the pursuit of more effective ways of understanding the physical world. It is in this context important to understand scientific knowledge (perhaps pertaining to the nature of climate change, or the health impacts of particulate air pollution for example) not as unchallengeable truths, but as the best available knowledge that we have at that particular moment in time. This is precisely why the Austro-British philosopher Karl Popper famously observed that it is the ability to be proven wrong (or 'falsified'), and not its relationship to essential truths, which marks scientific knowledge out from other forms of knowledge, such as those pursued in philosophy and the social sciences (Popper, 2002 [1950]).

Recognizing that it is the ability of scientific knowledge to be proven wrong – not its association with essential truths – marks it out as a distinctive form of knowledge, and enables us to appreciate the complex forces that shape and change the nature of scientific understandings. A crucial factor within the production of scientific knowledge is the notion of *scientific consensus*. At any given time

Box 3.4 Naomi Oreskes, 928 papers and the climate change consensus

In a 2004 paper published in the journal *Science*, and entitled, *The Scientific Consensus on Climate Change*, Naomi Oreskes argued that there was a clear and compelling consensus within the scientific community on climate change. Oreskes's research was inspired by the continued media depictions of the uncertainties surrounding the science of climate change. Oreskes's study looked at 928 peer-reviewed scientific papers that were published between 1993 and 2003. The study classified these papers on a grid ranging from 'explicit endorsement' of the position that recent observed patterns

Plate 3.4 Naomi Oreskes
Source: Wikimedia Creative Commons, Ragesoss

of warming in global average temperatures are a result of human-induced climate change, to the 'explicit rejection' of this position. Oreskes's study revealed that 75 per cent of the papers analysed either explicitly or implicitly endorsed the theory of human-induced climate change, and that none of the papers disagreed with this position.

Oreskes's intervention into the climate change debate raises some interesting questions about the nature of scientific consensus. For example, do the number of people involved in a consensus add to its strength, or should the strength of a consensus be measured by the quality of scientific research and theories that support it? It could be, for example, that the 928 papers reflect a kind of herd mentality, in and through which scientists follow the established studies and findings of a few, and thus artificially inflate the apparent strength of the consensus. Furthermore, Oreskes herself acknowledges that the consensus that she observes could be wrong. What is clear from Oreskes's study is that the uncertainties that surround the science of climate change appear to be more about its precise implications than the question of whether it is a product of human interventions in the Earth's atmosphere.

Key readings

Oreskes, N. (2004) 'The scientific consensus on climate change', *Science* 306: 1686
Oreskes, N. and Conway, E.M. (2010) *Merchants of Doubt*, Bloomsbury Press, London
Watch Oreskes talk about the scientific consensus on climate change at: http://www.youtube.com/watch?v=XXyTpY0NCp0 (accessed 5 July 2013)

there may be various scientific theories concerning any number of scientific questions (the impacts of passive smoking on human health; or the role of the Higgs Boson Particle in the fabric of existence, for example). In this context, scientific research progresses through the development of consensuses, in and through which particular ways of understanding the world become accepted wisdoms, while other theories are rejected or ignored. In 2004, for example, the science historian Naomi Oreskes proclaimed that despite the ongoing debate, there was a consensus within science that observed levels of climate change were a result of human activities (Oreskes, 2004). But, of course, history has consistently shown that often the scientists and theories that are rejected within scientific consensus turn out to be right all along.

In order to explain the changing nature of scientific knowledge, the American physicist Thomas Kuhn put forward the notion of *paradigm shifts*. According to Kuhn, a paradigm is a scientific model that explains how a certain process works. Paradigm shifts occur when two, incompatible, theories of scientific knowledge vie with each other to be the accepted way of explaining how the world works (see Kuhn, 1962) (see Figure 3.2). So a paradigm shift could be observed in geography, when theories of a flat Earth were finally rejected, or in astrophysics when Einstein's work challenged Newtonian theories of the universe. It is important to realize that Kuhn did not suggest that paradigm shifts are quick and easy transitions between different scientific models of the world. It is, perhaps, best to think of paradigm shifts as periods of crisis when competing scientific theories and scientists struggle to establish new scientific consensuses.

These discussions about the nature of science and scientific knowledge are important to our broader discussion of the Anthropocene for two reasons. First, because the modern scientific method, and the sense of human mastery over the environment it has often promoted, has been a central driving force in the production of a geological era defined by humans. Second, the

---

Box 3.5 Scientific consensus building and the pasteurization of France

The French philosopher Bruno Latour provides an interesting example of scientific consensus building in his book *The Pasteurization of France* (Latour, 1993). In this book, Latour describes how Pasteur's breakthroughs in how we understand microbiology, the spread of diseases and immunization were not generally accepted and implemented on the basis of the truth of his scientific studies. The pasteurization of France, and the new germ theories on which it was based, were the product of a long process of consensus building, which spanned microbiological laboratories, the offices of urban planners, the recommendations of doctors and a broader public hygiene movement. None of this is to say that the theories of Pasteur, and countless other prominent scientists, where not accurate or effective. But in recognizing the changing nature of scientific knowledge and the practices of consensus building, we can hopefully begin to understand the role that social circumstances and political forces can have in shaping what we know about the environment and how we act on that knowledge.

Key reading

Latour, B. (1993) *The Pasteurization of France*, Harvard University Press, London

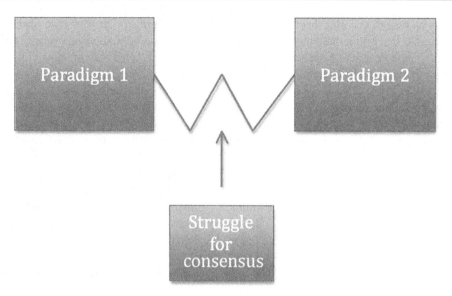

Figure 3.2 The nature of paradigm shifts
Source: Mark Whitehead

current discussions that are being orchestrated by the International Commission on Stratigraphy about the Anthropocene (see Chapter 1) remind us that this new geological era cannot simply come into existence: if it is scientifically recognized it will be the product of the building of a new scientific consensus in the geological community.

3.3.2 Atmospheric uncertainties and early atmospheric science

Before moving on to consider a geographical case study example of the relationship between atmospheric change and science, I want to reflect upon two scientific studies of the atmosphere, which collectively help to reveal the complex and contested nature of scientific knowledge. The first example takes us back to the early part of the twentieth century and emerging attempts to measure and quantify air pollution in British cities.

The quote that follows is an account (cited in Whitehead, 2009: 37) of 42-mile journey undertaken by a Mr Bloor around the city of London on 31 January 1925:

From 10am to 1pm visibility was good, but after that became misty making observation difficult. On the whole as far as smoke was concerned, the observation was disappointing, with hardly any of the numerous shafts emitting smoke during the time I was there. The following were some that were emitting smoke today. The square brick shaft at the Barking Gulford was the worse case of black smoke I saw today, close by lay a shaft owned by the Cape Asbestos Co emitting dense black smoke.

Mr Bloor was one of a growing number of smoke inspectors employed by London County Council to police air pollution in the metropolis. Smoke inspectors were expected to provide accurate, scientifically based accounts of air pollution events, upon which various legal prosecutions could be based. Putting ourselves in the shoes of Mr Bloor enables us to begin to understand some of the challenges that air pollution presents to science. First, as an air pollution inspector, Mr Bloor's ability to accurately assess pollution events was, in part, determined by the atmospheric conditions of the day. As we see from his reflections on 31 January 1925, as soon as mist descended his observation become compromised. Second, because Mr Bloor's version of atmospheric science had to be carried out at street level, his scientific objectivity became compromised at a series of levels. At one level, Mr Bloor could only observe a limited number of chimneys at one time, so his scientific gaze was obviously limited in its scope. His scientific gaze was also dependent upon the local terrain, and his ability to get a suitable vantage point from which to view air pollution. At another level, there are various accounts of smoke inspectors being approached by factory owners as they carried out their inspections, who would plead for leniency against prosecution. These factors serve to remind us that atmospheric science is rarely something that can be confined to a laboratory. As a science that must be practised in the midst of the complexities of everyday life, the study of air pollution must acknowledge the particular conditions that surround the knowledge that it produces.

In order to address the problems that faced British smoke inspectors, the national government and local authorities eventually promoted the use of various instruments for measuring pollution. While these gauges and filters could offer a more reliable and continuous assessment of urban air pollution, questions still remained about where they should be located (Whitehead, 2009: 94–125). Many air pollution instruments that were placed in city-centre locations were subject to vandalism (with some deposit gauges in Glasgow routinely containing urine following busy Saturday nights in the city). More recently, the 'urban bias' in the location of air pollution monitoring stations has become an issue within the broader politics of acid rain. According to Hajer (1995: 128), the decision to locate the majority of the UK's sulphur dioxide monitoring stations in urban areas led to gaps in scientific knowledge concerning the extent of acid rain deposition in rural areas. More worryingly, this lack of scientific knowledge appears to have led

to the false belief within the British government that acid rain was not having a significant impact on the ecologies of the countryside in the UK (once rural sulphur dioxide deposition monitoring did begin in rural areas, it was shown that levels of acid rain deposition in northern Britain were comparable with those in southern Sweden: a well known acid rain hotspot) (Hajer, 1995: 128–129). What these examples illustrate is that science must always be located somewhere, and this location has impacts on the types of scientific knowledge that are produced (see Haraway, 1991).

3.3.3 Climate change denial: from big tobacco to climategate

If the example of air pollution monitoring illustrates the practical conditions that limit the scope and validity of the scientific knowledge of the atmosphere, recent trends in climate science indicate how scientific knowledge can be more deliberately manipulated (see Hulme, 2009). While many scientists have raised important questions about the scientific consensus on human-induced climate change, others have sought to deliberately generate a sense of public confusion on the issue.

In his book *Heat: How to Stop the Planet Burning*, *The Guardian* journalist and prominent environmental author George Monbiot outlines what he terms the climate change 'denial industry' (Monbiot, 2006). As a way of illustrating the nature of this denial industry, Monbiot reflects upon a statistic that was reported by the prominent British botanist David Bellamy in a 2005 letter published in the *New Scientist* magazine. Bellamy's letter stated that the latest scientific research suggested that counter to the claims of the climate change consensus, 555 out of the world's 625 glaciers had been growing since 1980 (Monbiot, 2006: 24). When Monbiot asked where Bellamy had got this data from he identified the source as an issue of the journal *21st Century Science and Technology*. It was at this point that Monbiot become very suspicious about the source of Bellamy's unlikely data. Monbiot discovered that *21st Century Science and*

Technology belonged to the American millionaire Lyndon Larouche. According to Monbiot, Larouche is perhaps not the most reliable place to obtain your scientific data. In addition to casting doubt on human-induced climate change, Larouche holds some controversial views on various aspects of global political life (Monbiot, 2006: 25). While no credible scientific source can be found for this glacier data, it has been used by a range of institutions and individuals to cast doubt on the scientific consensus on climate change.

Monbiot did notice that many of the groups who were circulating this dubious data had something in common: they had received funding support from ExxonMobil.[5] On many measures ExxonMobil is classified as the world's largest corporation. It does, of course, depend for its continuing success on its ability to extract, refine and sell oil. On these terms ExxonMobil's commercial success is challenged by a scientific consensus that suggests that the continued use of fossil fuels could generate disastrous climatic consequences. It should come as little surprise that the scientific research that ExxonMobil funds often casts doubt on the climate change consensus and the science upon which it is based. Sim (2006) has described this type of science as a form of 'special interest scepticism'. Given the overwhelming breadth of scientific evidence that supports the twin notions that climate change is occurring and that this change is caused by human interventions in the global atmosphere, it is difficult for climate change sceptics to produce scientific evidence that is able to prove this consensus wrong (unless fabricated evidence, such as the previously mentioned glacier statistics, can be circulated). Monbiot argues that in these circumstances organizations such as ExxonMobil, and the scientists and right-wing politicians that they support, have not sought to prove the scientific consensus wrong. Instead they have worked to suggest that the science of climate change is marked by great uncertainty and division. This strategy is perhaps best encapsulated in the words of Frank Lunz (cited in Monbiot, 2006: 27), who acted as a political consultant for George W. Bush, when he stated (in a leaked memo):

> Should the public come to believe that the scientific issues are settled, their views about global warming will change accordingly. Therefore you need to continue to make the lack of scientific certainty a primary issue in the debate.

While discussing the genuine scientific uncertainties that remain about climate change is a healthy thing to do, it appears that many climate change deniers are deliberately manufacturing uncertainty in order to generate a confused understanding of climate change in the minds of the general public. It appears that climate change deniers have found an unlikely source of inspiration in their manufacturing of scientific uncertainty. The tobacco industry has been fighting the findings of mainstream science for some time. In order to avoid compensation claims and governmental regulation, so-called 'big tobacco' has been funding scientific research that has routinely cast doubt on the link between smoking, passive smoking and lung cancer. While is clear that those wishing to cast doubt on the climate change consensus have learnt a lot from the endeavours of the tobacco industry, Monbiot suggests that there is now an alliance emerging between the two groups (2006: 20–39). This alliance is based upon a recognition that through carefully coordinated scientific research programmes and publicity campaigns, big oil and big tobacco can collectively work to cast doubt over the findings and intent of mainstream science, and thus ward off unwanted governmental regulations in the fields of smoking and climate change.

More recently, scientific controversies around climate change took an unexpected turn. In 2009, just before the United Nations Conference on Climate Change in Copenhagen, a file containing 1000 emails sent from and received by members of the prominent Climate Research Unit at the

Box 3.6 Bjørn Lomborg – the sceptical environmentalist

One attempt to develop a serious discussion about the nature of climate science and the threats of climate change was instigated by the Danish academic Bjørn Lomborg. In his 2001 book *The Skeptical Environmentalist*, Lomborg cast doubt on various scientific predictions that suggest impending ecological doom for humanity. In his reflections on climate change, Lomborg accepted that climate change was happening and that it was caused by humans. He did, however, question predictions that climate change would lead to serious socio-ecological problems in the near future. Lomborg thus argued that there were more pressing issues facing humanity than climate change, and that these should be given a higher priority.

Lomborg's book proved to be controversial and resulted in his work being investigated by the Danish Committees on Scientific Dishonesty. In 2003 the Danish Committees on Scientific Dishonesty ruled that *The Skeptical Environmentalist* did exhibit scientific dishonesty, but that this dishonesty was not a result of gross negligence or a deliberate intent to mislead on the part of Lomborg. In essence the Danish Committees on Scientific Dishonesty felt that the dishonesty evident in Lomborg's book was largely a product of his lack of scientific expertise in the field of climate change. Given that climate change embraces sciences as diverse as glaciology, hydrology, climate science and meteorology, *inter alia*, this is a charge that could be levelled at many working in the field of climate change studies. *The Economist* magazine defended Lomborg's work, suggesting that his book never claimed to be a scientific study and should not thus have been subject to a scientific dishonesty hearing. What the case of Lomborg, and his *Skeptical Environmentalist* book, clearly illustrate is that climate change raises questions about the nature of scientific expertise and dishonesty.

Key readings

Economist (2003) 'Thought control', *The Economist*, 9 January
Lomborg, B. (2001) *The Skeptical Environmentalist*, Oxford University Press, Oxford

University of East Anglia were leaked onto the internet (see Nerlich, 2010). Although these emails had been obtained through the illegal hacking of scientists' email servers, they raised questions about the impartiality of well-respected climate scientists, and led to what became know as Climategate. For a long period of time scientists working at the Climate Research Unit had been strong supporters of the climate change consensus and had produced reliable data to support the case for anthropogenic climate change. The intercepted emails related to correspondences that had been undertaken over a 15-year period. Climate sceptics took a particular interest in emails that discussed attempts to stop the publication of

certain scientists' work, and how data could be adjusted using certain 'tricks' (Nerlich, 2010: 422). Although an independent panel cleared the Climate Research Unit of 'scientific impropriety or dishonesty', Climategate revealed how easy it is to discredit scientific consensuses and undermine

For a discussion of the role of the media in shaping the climate change debate and the ways in which climate science is interpreted by the public go to *The Yale Forum on Climate and the Media*:

http://www.yaleclimatemediaforum.org/

the policy regimes that have been built upon them. It is crucial in this context to be open about the inevitable human bias that surrounds scientific research, but not to equate these biases with bad science and the dishonest practices of special interest sciences.

3.4 CORRIDORS OF UNCERTAINTY: 'FUGITIVE EMISSIONS' AND THE CASE OF LOUISIANA'S CANCER ALLEY

This section considers the complex connections that exist between science and atmospheric pollution by reflecting on recent events in a particular place. The place in question is the community of Diamond in Louisiana's so-called chemical corridor. As we will see, this geographical community has, in many ways, been at the forefront of evolving relationships between science and atmospheric pollution in the Anthropocene.

3.4.1 Diamond: an anatomy of a polluted place

Diamond is a small community, comprising of only four streets, which is located in the small town of Norco. Norco is on the northern banks of the Mississippi River between New Orleans (to the east) and Baton Rouge (to the northwest) (see Figure 3.3). The region between New Orleans and Baton Rouge is often referred to as a 'chemical corridor' due to the high number of oil refineries and chemical plants that are located around the Mississippi River (and which are supplied by the crude oil that arrives by tankers from the Gulf of Mexico). Even in a region that is renowned for environmental pollution, the community of Diamond is located in an unfortunate place.

The small community of Diamond sits in close proximity to a chemical plant and an oil refinery. In his fascinating study of Diamond, Steve Lerner (2005: 9) describes the situation in the following terms:

Diamond is not a place where most people would choose to live ... the four streets of this subdivision are hard up against fencelines of a Shell Chemical plant and the huge Shell/Motiva oil refinery. Residents have long breathed the fumes from these two plants, suffered illnesses they attribute to toxic exposures, and mourned neighbours and friends killed by explosions at these facilities.

In many ways, Diamond reflects a significant point of convergence for the varied forces that have contributed to the transformation of the atmosphere in the Anthropocene. At the one level, in the production of various chemicals that are derived from oil, Diamond experiences the local release of a heady mix of air pollutants (including VOCs and chemical smogs), which have become synonymous with modern air pollution. At another level, the refined oil, produced in and around Diamond, supplies the cars and power plants that emit troubling quantities of carbon dioxide into the global atmosphere.

Shell's presence in the area of Diamond can be traced back to 1916 when the New Orleans Refining Company (where Norco gets its abbreviated name from) purchased 366 acres of cane field in the area and established a petroleum supply terminal there. The New Orleans Refining Company was a subsidiary of the Shell Corporation, and it commenced the refining of oil on this newly purchased site in 1920. In the 1950s Shell purchased more land in this area so that it could begin the construction of a new chemical plant. The land that Shell purchased was, at the time, occupied by the descendants of freed slaves, whose families had lived in the area since the end of American civil war (Lerner, 2005: 12). These families rebuilt their homes on the site of what would become the modern-day community of Diamond. While these displaced communities moved to the fence lines of Shell's petrochemical plants in the hope of gaining employment, over time they became the frontline recipients of a

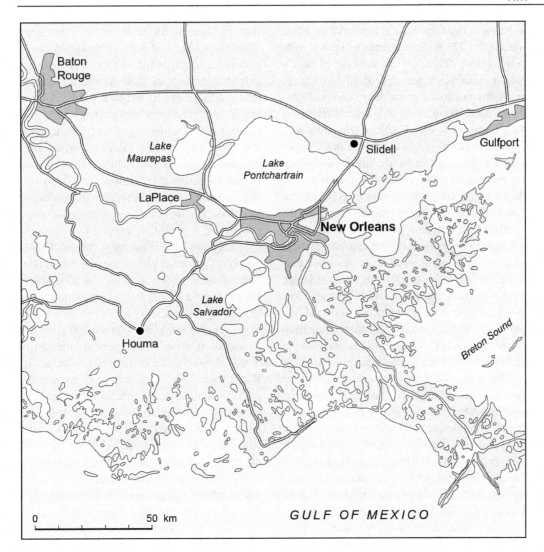

Figure 3.3 Chemical corridor, Louisiana
Source: Mark Whitehead

cocktail of air pollutants that were emitted from Shell's refinery and chemical plant (Lerner, 2005).

3.4.2 Air science, fugitive emissions and corporate liabilities

The case of Diamond raises a series of issues concerning corporate environmental responsibility (see Chapter 5), vulnerability and environmental injustice (see Chapter 9) and resource use

(see Chapter 2) that are discussed elsewhere in this book. In this chapter, I want us to focus on one particular aspect of the story of Diamond: the connections it exposes between science and atmospheric pollution.

The levels of air pollution in Diamond, and the broader industrial region within which it is located, have generated problems for scientists trying to assess and monitor the quality of the area's air. In the 'chemical corridor that runs from New Orleans to Baton Rouge there are 156 industrial facilities, which emit 129 million pounds of toxins every year' (Lerner, 2005: 44). In the case of Shell's Norco refinery, 75 per cent of all of its toxic emissions go into the atmosphere (Lerner, 2005). Unintended air pollution in Diamond takes two basic forms: 1) so-called 'fugitive emissions' that emanate from leaky pipes and valves; and 2) larger scale accidental release events. Fugitive emissions are particularly problematic to human health as they tend to come from sources close to the ground, where they are more likely to come into contact with humans (as opposed to those emission that are released from chimneys, which enables air pollution to dissipate at higher altitudes (Lerner, 2005: 44–52)). In terms of larger scale accidental releases, between January 1990 and September 2000 Shell officially reported 341 pollution 'events', which released dangerous chemicals such as the carcinogen benzene into the air around the Norco plants (Lerner, 2005: 48).

The scale and nature of air pollution in Louisiana's chemical corridor generates scientific problems at a series of levels. First, the sheer scale of air pollution in the region means that it is impossible for scientists and regulators to monitor all air pollution events. The work of scientists is also hindered by the clandestine activities of corporations such as Shell, who, it is claimed, use the cover of mist, fog and darkness to hide their air polluting activities (Lerner, 2005: 44–52). Scientists working for chemical corporations in the region also use the close proximities of chemical plants as a way of shifting the blame for air pollution from their own companies on to other plants. While corporations are expected to keep their own inventories of toxic air pollution releases, the leaks and associated forms of fugitive air pollution tend to go unrecorded and unmonitored (estimates suggest that in Diamond some 80 million pounds of VOC emissions go unreported due to leaks every year) (Lerner, 2005: 44).

Beyond the practical challenges facing atmospheric scientists working in Louisiana's chemical corridor, there are other scientific issues that impinge upon their work. Even when definitive evidence of toxic air pollution can be gathered and attributed to particular polluters, corporations such as Shell dispute the impacts that pollution events have on local residents. While local residents complain of sore eyes, respiratory difficulties, reduced fertility rates and cancers that they attribute to breathing in polluted air, the complex nature of these illnesses (which are connected to genetics and lifestyles as well as environmental factors) makes it difficult for medical scientists to provide causal proof connecting a pollution event to a particular illness. Perhaps more worrying still, while Shell openly acknowledged the release of some 2 million tons of toxic chemicals into the atmosphere around Diamond in 1997, the corporation claims that these releases still fall within permissible environmental guidelines (Lerner, 2005: 47). The issue that is raised here concerns the setting of safe thresholds for permissible pollution (see Whitehead, 2009). The setting of permissible pollution standards involves close collaborations between governmental officials and scientists, but often reveals uncertainties within the scientific community concerning precisely what safe levels of exposure to toxic pollution are. In the context of the demands for economic development and job creation in places such as Diamond, there is a real danger that even if air pollution could be effectively measured and monitored that acceptable thresholds for air pollution emissions would still be set at too high a level.

Many in the community of Diamond are now seeking relocation away from Shell's Norco complex. In order to achieve this they are requesting financial compensation and support from Shell. Given the uncertainties that surround the levels of air pollution emanating from Shell's refinery and chemical plant, the impacts of these forms of pollution on human health, and what constitutes acceptable levels of pollution, such compensation may be difficult to achieve. It is in the context of situations such as those found in Diamond that the potential and limitations of science in regulating socio-atmospheric relations becomes apparent. While science can provide a sure basis for regulating and penalizing polluters (Shell's Norco complex has been subject to a series of financial penalties), its requirement for definitive evidence can also be a roadblock to protecting communities and broader ecosystems from long-term harm. It is in situations like these that a mix of science and common sense can provide a useful solution to atmospheric controversies.

3.5 CONCLUSIONS

In this chapter we have explored the ways in which the nature of the atmosphere has been shaped by human activity in the Anthropocene. As part of this process we have seen how human interventions in the atmosphere have contributed to changes in the content of the atmosphere (including acid rain and photochemical smogs) and in how the atmosphere as an environmental system operates (in relation to climate change and ozone depletion in the upper atmosphere). The account of atmospheric change in this chapter has been complemented by a consideration of the role of science in both contributing to the transformation of the atmosphere and in helping humankind to better understand the nature and extent of these changes.

Our discussions of scientists have ranged from individuals such as Thomas Midgley and Charles Keeling, to a broader consideration of the nature of modern science itself. By considering the broader role of science in atmospheric affairs we have seen that while science can often provide a sound basis for judging the extent and impact of atmospheric change, care must be taken when using scientific evidence. Although science is often associated with providing answers, it is important to realize that these answers are often partial approximations. Scientific theories and knowledge change over time as established consensuses rise and fall. Various practicalities can often hinder the ability of science to develop definitive pictures of the nature of human-induced atmospheric

change. Nowhere are these issues illustrated more effectively than in the field of climate change. While there is a scientific consensus that climate change is happening and that it is caused by human activities, there remains considerable uncertainty about the precise impacts that climate change will have on the planet. Furthermore, so-called 'special interest' science, which is often funded by large corporations, is deliberately spreading misleading information that is obscuring the truth about climate change. Having read this chapter you will hopefully now be in a position to understand the crucial role that science plays in helping society deal with the environmental issues that are emerging in the Anthropocene. This chapter should, however, remind you to be constantly vigilant when you assess the validity of scientific evidence and be conscious of the conditions under which such evidence has been produced.

NOTES

1 This occurs when pockets of air and fuel combust at suboptimal times within the internal combustion engine.
2 Early fridges utilized toxic gases such as methyl chloride, which were highly dangerous when leaks occurred.
3 The annual fluctuations in atmospheric carbon dioxide identified within the Keeling Curve reflect the fact that there is an annual increase and decrease in global CO_2. This fluctuation is the product of the fact that when trees and foliage grow in summer seasons they absorb carbon dioxide from the atmosphere (reducing its overall concentration); in the autumn this CO_2 is returned to the global climate (increasing atmospheric concentration of carbon dioxide). The reason that this fluctuation is not smoothed over by the counterbalancing of southern and northern hemisphere summers and winters is because the northern hemisphere has more land mass (and plant life) than the south and thus generates a globally significant seasonal flux in carbon dioxide levels.
4 The full text of *Fumifugium* can be downloaded at: http://ia600204.us.archive.org/6/items/fumifugium00eveluoft/fumifugium00eveluoft.pdf (accessed 3 August 2012).

5 Monbiot did not, however, claim that David Bellamy had received funding from ExxonMobil.

KEY READINGS

Hulme, M. (2009) *Why We Disagree About Climate Change: Understanding Controversy, Inaction and Opportunity*, Cambridge University Press, Cambridge. This book provides an excellent overview of the nature of the scientific debate within the field of climate change studies.

McEwan, I. (2011) *Solar*, Vintage Books, London. This novel provides a very interesting and humorous account of the role of science in the field of climate change. The book tells the story of the Nobel Prize-winning physicist Michael Beard and his journey from being indifferent to the threats of climate change to becoming one of the leading figures in the pursuit of alternative energy sources that could be used to avert climate catastrophe.

McNeill, J. (2000) *Something New Under the Sun: An Environmental History of the Twentieth Century*, Penguin, London. This volume provides a comprehensive and highly accessible account of issues relating to atmospheric pollution and climate change.

CHAPTER FOUR

Soil

The political ecology of soil degradation

4.1 INTRODUCTION: GETTING UNDER THE PLANET'S SKIN

Have you ever wondered what it would be like to live in a world without soil? For eight years during the 1930s those living on the Great Plains of the USA and Canada got a chilling insight into just what such a world might look like. The Great Plains of North America (referred to as the Prairie in Canada) make up one of the most significant agricultural regions in the world. Covering an area of some 400 million acres that runs through the heart of the North America continent (see Figure 4.1), the flat relief and fertile soils of this region have made it highly conducive to a range of agricultural practices. The most significant crop grown on the Great Plains is wheat, but the region is also synonymous with the cultivation of barley, corn, cotton and soybeans (among many other things) (Hudson, 2011). In addition to supporting the growing of food crops, the Plains are also home to vast ranches that are devoted to the rearing of sheep and cattle.

Throughout the history of the US and Canada, the Great Plains have played a vital role in feeding the rapidly expanding populations of these countries. But during the 1930s something happened to disrupt the ecological balance of this region. During this time period a vast dust bowl developed in the southern section of the Plains. The environmental historian Donald Worster

(1979) claims the Dust Bowl was one of the most devastating human-induced ecological disasters in history. As with many of the cases of soil erosion we discuss in this chapter, the Dust Bowl was the product of a number of environmental and social factors. At one level, as you would expect, the Dust Bowl was partly caused by an enduring drought that lasted for much of the 1930s. But the Dust Bowl was also a product of the agricultural practices that had become established in the region. In the pursuit of the great profits that could be made from the growing of wheat, farmers engaged in ploughing up vast swathes of the natural grasslands found in the Prairies. The so-called 'Great Plow Up' left the soils of the southern Great Plains exposed to the processes of soil erosion in ways they had never been before. When drought and strong winds came, vast swathes of the fertile topsoil of the Plains was removed. This process of soil erosion had two main consequences. First, it made it very difficult to produce agricultural crops in the region, thus consigning large numbers of agricultural families to poverty and hunger. Second, the Dust Bowl became associated with vast dust storms (comprised of the soil that had been removed from the land). These storms, or 'blackouts' as they were often called, would regularly block out the sun (thus impairing agricultural productivity in the region) and cause respiratory problems among the resident population. Given these strange, almost otherworldly

> ## A history of the Great Plains: *The Plow that Broke the Plains*
>
> The 1936 documentary film *The Plow that Broke the Plains* provides a helpful account of the varied processes associated with the emergence of agriculture on the North American Prairie. The documentary also explains how the Dust Bowl emerged and the socio-ecological consequences that were associated with it.
>
> In YouTube, search for *The Plow that Broke the Plains* or go to: http://www.youtube.com/watch?v=fQCwhjWNcH8 (accessed 7 January 2013).

conditions, it is unsurprising to learn that the Dust Bowl resulted in the mass migration of people away from the Great Plains. In essence, a society without its soil is quite simply no society at all.

In many respects, the story of the Dust Bowl is one of the defining moments in the history of the Anthropocene: a moment when the capacity of humans to disrupt soil systems on an unprecedented scale became apparent. In this chapter we explore the changing nature of human relations with soil. While soil is a complex compound that takes many forms, this chapter understands it as, '[T]he biologically active, porous medium that has developed in the uppermost layer of the Earth's crust' (Merriam Webster, 2012). On these terms, it is helpful to think of soil as a form of *living skin*, which coats large parts of the planet. As a living skin, soil provides a collective home for water, nutrients and organic matter. It also provides a key link in both the carbon and nitrogen cycle, as these two vital elements circulate around the Earth's ecosystems. In addition to distributing key nutrients, soils also provide a series of so-called ecosystem services. Ecosystem services are the valuable forms of assistance that things like soil constantly provide to the broader biological communities of which they are a part. In this context, soils act a kind of filter through which water can be cleaned and reused within an ecosystem. As a form of interface zone between life and death, soils also enable decaying waste matter to be broken down and reused within an ecosystem.

This chapter argues that the human use and exploitation of soils has been a defining characteristic of the Anthropocene. As humans have sought to utilize soil resources in order to feed an expanding global population, the condition and quality of soils throughout the world has been transformed. Some of these transformations have been regional in nature, as topsoils have been over exploited and eroded. Human attempts to enhance the quality of soils through the application of artificial fertilizers have, however, also had a significant impact on the global nitrogen cycle and the varied ecosystems that depend upon it. This chapter begins by charting the human impacts on soil and the nitrogen cycle. The second section introduces the work of political ecologists, who have developed helpful frameworks in and through which it is possible to understand the transformation of soils as the complex outcome of physical environmental, political, social and economic factors. The final section of this chapter explores all of these themes in greater detail through the case of modern soil erosion in China.

4.2 SOIL AND ENVIRONMENTAL TRANSFORMATIONS

4.2.1 The human colonization of soil: a brief history

The large-scale human transformation of soils began some 12,000 years ago. It was at this point

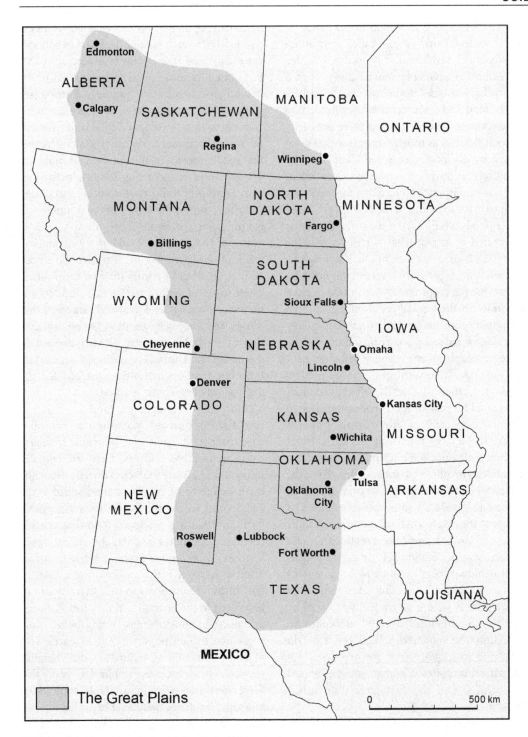

Figure 4.1 The Great Plains of North America
Source: Centre for Great Plains Study

that we start to see a transition in the nature of human society. During the so-called agricultural (or Neolithic) revolution that occurred at this time, humans gradually moved away from a society that was based upon hunting and gathering the food and resources that they needed, to an agricultural type of society. There are many different theories as to why this great change in human society occurred. Some claim that the transition may have been a fortunate accident, as early hunter-gatherers noticed that the seeds they dropped in the soils around their camps produced the very foodstuffs they had travel so far to procure. However it occurred, what is clear is that the adoption of agricultural techniques had significant advantages for early human communities. By growing the crops that they required in close proximity to their places of dwelling, human communities no longer had to expend energy searching for naturally occurring foodstuffs. The domestication of plants (namely bringing plant growing under human control) was followed by the domestication of animals, whereby animals were reared for milk or food production close to human settlements. In addition to making human life far less energy intensive, the agricultural revolution also made life far more secure. This security can be observed at two levels. First, the deliberate cultivation of crops and animals enabled humans to develop a more reliable source of food: with the vagaries of finding and successfully hunting food over long distances replaced by more reliable harvests. Second, agricultural techniques enabled human communities to become geographically settled. Migrating communities of hunter gathers were thus gradually replaced by sedentary communities that could offer collective protection and support for their residents (in Chapter 6 we explore how the new securities that agriculture offered human society enabled the formation of the first large-scale urban settlements).

The new found securities associated with agricultural society enabled human communities to flourish and resulted in significant increases in human population. As human populations expanded, however, it became necessary to cultivate expanded tracts of soil. Some 6000 years ago (4000 BCE), the increasing demand for soil led to the first large-scale forest clearances in Europe (see Chapter 5). Inevitably, however, there are limits to the amount of new agricultural lands that can be created by forest clearances. It is in this context that modern forms of agriculture have sought to develop ways of increasing the productivity of soils. During the industrial revolution that started in Britain during the eighteenth and nineteenth centuries, agriculturalists became increasingly aware of the benefits of adding inorganic and artificial fertilizers to soils. Farmers have been adding fertilizers to soils for many centuries. These so-called organic fertilizers include animal manure, compost and seaweed, among other things, and essentially result in the natural cycling of nutrients that are already present in the biosphere. During the industrial revolution, however, new opportunities emerged for the application of inorganic fertilizers.

Inorganic fertilizers are minerals that are comprised of nitrogen, potassium or phosphorous compounds, *inter alia*. Inorganic fertilizers come in two basic forms. First, are 'mined' minerals, which are extracted directly from the earth, before being processed and applied to the soil. Second, are chemically synthesized inorganic fertilizers that are produced following specific chemical procedures, such as the Haber–Bosch Process. What unites both of these forms of artificial fertilizer is that unlike organic sources, they introduce nutrients to soils that would not have been in the ecosystem if they had not been introduced by human beings. During the industrial revolution in Europe, significant advances were made in both mining techniques and chemical sciences. These processes combined to ensure that there were more readily available supplies of inorganic fertilizers that could be added to soils. Recent estimates claim that the last 50 years have seen an unprecedented rise in the application of inorganic fertilizers to global soils (Glass, 2003).

Between the 1950s and early years of the 2000s, it is claimed that inorganic fertilizer use rose 20-fold, with approximately 100 billion kilograms of nitrogen fertilizers being added to soils per annum (Glass, 2003). The use of inorganic fertilizers has been associated with the rapid expansion of crop yields and global food supply (see Plate 4.1). As we discuss in section 4.2.3 below, however, the rising use of inorganic fertilizers has led to a series of unintended environmental problems.

The remainder of this section considers the nature and extent of the human transformation of soil. It does so by focusing on the physical degradation of soils before moving on to consider the questions of soil pollution.

4.2.2 'Busting the Crust': soil degradation

The physical degradation of soil (such as the previously discussed case of the Dust Bowl) is one of the most pressing environmental issues of our times. The Food and Agriculture Organization of the United Nations (FAO) defines soil (and

Plate 4.1 The promotion of inorganic fertilizer use by the Tennessee Valley Authority
Source: Wikimedia Creative Commons

more generally land) degradation as, 'the reduction in the capacity of the land to provide ecosystem goods and services and assure its functions over a period of time for the beneficiaries of these' (FAO, 2012). According to the FAO, soil and

---

Box 4.1 Haber–Bosch Process

Described by Smil as the most important invention of the twentieth century, the Haber–Bosch Process was developed by the Germany scientists Fritz Haber and Carl Bosch. The process itself enabled the efficient production of ammonia (a key source of nitrogen) from the chemical synthesis of hydrogen and nitrogen. Various compounds of ammonia could then be added to soils in order to improve their fertility. It has been claimed that the Haber–Bosch Process, and its impact on global food production, was the most important factor in the global population explosion of the twentieth century. Its significance can also be discerned in the fact that both Fritz Haber and Carl Bosch received Nobel Prizes for their roles in developing the process. Along with the development of the internal combustion engine, and chlorofluorocarbons (discussed earlier in this volume), it can be argued that the development of the Haber–Bosch Process was one of the defining moments in the emergence of the Anthropocene.

Key reading

Smil, V. (2004) *Enriching the Earth: Fritz Haber, Carl Bosch, and the Transformation of World Food Production*, MIT Press, Cambridge

land degradation costs approximately $40 billion every year throughout the world as a consequence of lost soil productivity.

Soil degradation involves a loss of organic matter, nutrients and water from land. There are various processes that contribute to soil degradation. It can, for example, be the result of forest clearances (see Chapter 5) in and through which soils get cut off from the water, carbon and nutrients that trees supply. Soil degradation is also often the result of the over use and poor management of land within modern agriculture. The overgrazing of pastures and the over-cultivation of arable land results in the net loss of nutrients and water from the soil and a decline in the ability of the soil to support future farming endeavours. In the case of forest clearances, water (in the form of rainfall) transfers nutrients away from soils (a process referred to as leaching). In the case of bad agricultural management practices, it is the crops and animals that remove nutrients from the soil. Studies indicate that imbalances in soil nutrient budgets are much higher in less economically developed countries than they are in more economically developed countries (agriculturalists in the latter appear much more able to apply inorganic fertilizers to their land). A study of 38 countries in Sub-Saharan Africa, for example, indicates that between 1983 and 2000 each country had a negative soil nutrient budget (with more nutrients being taken from soils than was returned annually) (Dent et al, 2007: 97). This means that some 950,000km² of land in Sub-Saharan Africa is under serious threat of long-term land degradation (Dent et al, 2007). The United Nations Environment Programme and the FAO now compile data on global levels of soil and land degradation. This research programme is called the Land Degradation Assessment in Drylands (LADA for short). This study, which is largely based on remote sensing satellite data, indicates that over the last 25 years there has been a 12 per cent decline in the net primary productivity (NPP) of land.

For more information on the UN LADA programme and its assessment reports go to:

http://www.fao.org/nr/lada/index.php

In the most extreme of cases, forest clearances and poor land management practices can result in soil erosion, or the net loss of topsoil from the land. Soil erosion is, of course, a natural process, but as we have seen in the case of the Dust Bowl it can be greatly accelerated by humans. Soil erosion occurs when exposed soil is moved by wind and/or water. While soil erosion is a global problem, it particularly afflicts dry land agricultural regions such as those found within Africa, Australia and parts of Asia. In West Africa, for example, estimates indicate that wind-based soil erosion affects 1.45 million km² of land (Dent et al, 2007: 95). Global estimates suggest that soil erosion results in the loss of 20,000 to 40,000km² of previously productive land.

The scale of contemporary patterns of soil degradation, expressed through both net nutrient loss and soil erosion, make it one of the defining environmental characteristics of the Anthropocene. But the effects of soil degradation are not evenly distributed in geographical terms: with dryland agricultural communities in less economically developed countries often being the most severely affected. The degradation of soil has a series of local and global consequences. At a local level, soil degradation is associated with the onset of human poverty. The large-scale loss of productive soil also results in local declines in rates of biodiversity, as the varied species that live in and depend on soil suffer as a consequence of its deterioration. In a global context, soil degradation is also a contributory factor in accelerating patterns of climate change. Soil degradation contributes to climate change in two main ways. First, soil erosion results in the release of carbon (from soil's organic matter) back into the atmosphere. Second, degraded soils are less able to support the trees and plants that can remove carbon from the atmosphere (recent estimates show that soil degradation resulted in 800 million tons less carbon being fixed from the atmosphere between 1981 and 2003 (Dent et al, 2007: 93)).

4.2.3 Soil pollution and the disruption of the nitrogen cycle

Soil degradation is the product of the unsustainable removal of water, nutrients and organic content from soil. There are, however, other environmental problems that are created by what is added to soil. Soil pollution can take two forms: 1) the unintended substances that are added to soil as a consequence of pollution events; and 2) the substances that are added to soil in order to enhance its productive capacity. In the first instance, toxic chemicals produced from various industrial processes routinely make their way into soil. Heavy metals, such as lead, have accumulated in soils over many decades largely as a result of the use of lead in petrol (see Chapter 3 for more details). Toxic chemicals such as mercury and cyanide (which are by-products of the mining and manufacturing industries) are transferred to soils through watercourses and drainage systems (Dent et al, 2007). Certain soils are now also home to the hydrocarbons that are the by-products of the petroleum industry, and to the radioactive substances that have been released by the nuclear industry. The accumulation of these harmful substances significantly reduces the ability of soils to support the ecosystems of which they are a part. Soil pollution of this kind also means that many areas can no longer be used to safely produce food for human production. This process places greater

---

Box 4.2 Rachel Carson

For many people, the publication in 1962 of Rachel Carson's *Silent Spring* was a defining moment in the emergence of the modern environmental movement. Carson was born in Pennsylvania in 1907 and went on to study English and science at Johns Hopkins University in Baltimore. Following her academic studies, Carson went on to become a biologist with the US Fish and Wildlife Service (Corcoran, 2001). Her love of English and biology led her to write a series of books on the sea: *Under the Sea-Wind* (1951); *The Sea Around Us* (1951); and *The Edge of the Sea* (1955) (Corcaran, 2001: 195). Carson's interest in the impacts of the application of DDT began in 1958 when she wrote an article exploring the impacts of the insecticide on bird populations (although it appears that she was concerned about the ecological impacts of the chemical back in 1945) (Corcoran, 2001). This article was a precursor to an extended period of research by Carson on the impacts of DDT, which ultimately led to the publication of *Silent Spring*. One of the most notable features of *Silent Spring* was its ability to combine the authority of science with a poetic concern with the natural world. Carson's work would ultimately lead to much tighter regulations on the use of pesticides throughout the world. Beyond this, however, Rachel Carson appears to be among the first scientists to recognize the scale of the environmental changes that humans were capable of achieving. As such, her work is a harbinger of the debates that we are now having about the Anthropocene.

Key readings

Carson, R. (1962) *Silent Spring*, Houghton-Mufflin Company, Boston, MA
Corcoran, P. B. (2001) 'Rachel Carson' in Palmer, J.A. (ed.) *Fifty Key Thinkers on the Environment*, Routledge, London: 194–200

pressures on the soils of other regions, thus increasing the likelihood of soil degradation there.

The second set of pollution processes that are associated with soils result from the deliberate application of pesticides and fertilizers to soils. Arguably the most infamous example of soil pollution resulting from the application of pesticides is the case of DDT. During the post-war period, DDT was a popular agricultural chemical that was used to control insects that would attack crops. Over time it was realized that DDT was having a harmful affect on the ecosystems within which it was being applied. In a famous book entitled *Silent Spring*, the American biologist Rachel Carson (1962) documented how once DDT had entered the environment through soils it was being rapidly transferred to plants and animals. As a so-called persistent organic pollutant (POP), DDT built up in the tissues of the wildlife that lived in the areas where the chemical was being applied. This chemical build-up resulted in a range of ailments that decimated local bird populations (among other species) and resulted in springs that were eerily silent due to the absence of birdsong.

Environmental problems can also be caused by the excessive application of artificial fertilizers to soils. As we have mentioned already, the application of artificial nitrogen and phosphorous-based fertilizers has enabled agriculturalists to greatly enhance the productive capacity of soils, and, at times, prevent soil degradation. But as levels of artificial fertilizer use have increased, it has become increasingly common for excess nitrogen and phosphorous compounds to be transferred

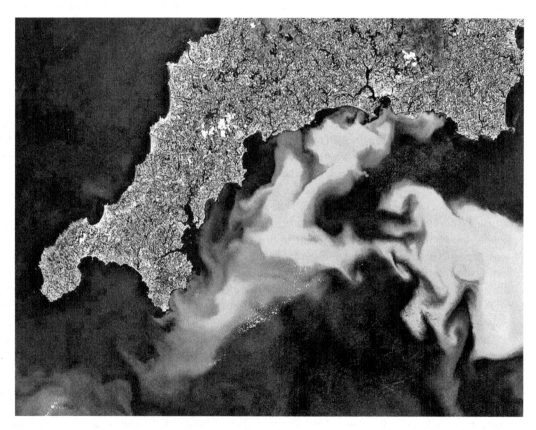

Plate 4.2 An algae bloom off the south coast of England
Source: Wikimedia Creative Commons, NASA

from soils into surrounding river systems and eventually the oceans. In rivers and oceans artificial fertilizers no longer support the growth of food crops but are instead absorbed by phytoplankton (more commonly referred to as microalgae). In warmer climates (particularly in springtime) the flow of nitrogen and phosphorous into the oceans can prompt the growth of huge algae blooms (see Plate 4.2). In the worse cases, these blooms can be so thick that they prevent light reaching the sea floor, creating 'dead zones' within the ocean.

The adding of nitrogen to soils represents one of the key ways that humans have been shifting the operation of global environmental systems during the Anthropocene. The forcing of the nitrogen cycle now means that in European states somewhere in the region of 70–75 per cent of all nitrogen that is fixed in soils is now added through artificial means (Dent et al, 2007: 100). The impacts of the addition of these levels of nitrogen to the global nitrogen cycle means that elevated levels of nitrogen can now be detected in underground aquifers and high-altitude clouds (Dent et al, 2007). The heighted presence of nitrogen in the global environment not only results in harmful algae blooms. When nitrogen is converted into nitrous oxides it actively contributes to both the erosion of the planet's ozone layer and to the production of acid rains. Notwithstanding the environmental problems that are associated with the anthropogenic boosting of the nitrogen cycle, large parts of the world now depend on

---

Dirt! The Movie

The recent documentary film *Dirt! The Movie* provides a helpful overview of both the role of soil in supporting life on Earth and the extent of the human transformation of soil resources. See:

http://www.youtube.com/watch?v=7Tj 9h18nrN4

---

nitrogen fertilizers for their food supply. As with so many problems that are associated with the Anthropocene, there is thus no easy solution to the current imbalance that we are seeing in the nitrogen cycle.

4.3 INTERPRETING THE TRANSFORMATION OF SOIL: A POLITICAL ECOLOGY PERSPECTIVE

So far this chapter has established the nature and extent of contemporary patterns of soil degradation and pollution. This section introduces *political ecology* as a framework within which we can better understand our collective relations with soil. Although notions of political ecology developed specifically in the context of analyses of land degradation, we see that the ideas presented within this body of work have more general application within the study of environmental change in the Anthropocene.

4.3.1 Piers Blaikie and the early political ecologists

As its name suggests, political ecology denotes a field of study that is interested in the connections that exist between social systems, biological processes, technological change, ecosystems, economic power and environmental transformation (see Bryant, 1998). While the term itself was first coined in the 1930s, it started to be used more widely during the 1970s (see Wolf, 1972). The notion of a political ecology perspective – at least as we understand it today – actually can be traced back to a study of socio-economic development in the Himalayan state of Nepal (see Simon, 2007: 699). The study was commissioned by the UK government's Overseas Development Administration and carried out by the British academics Piers Blaikie, John Cameron and David Seddon. The final report of this study, *Nepal in Crisis*, deployed a broadly Marxist perspective on the effects of international development (Blaikie et al, 1980).

This report was controversial (it was banned from being published for two years by the British government) because it argued that not only did international aid not address poverty and disadvantage in the developing world, but that it actively contributed to the perpetuation of these conditions (see Simon, 2007: 699). Blaikie and his colleagues argued that poverty in Nepal was the product of the systematic exploitation of marginal groups by ruling elites and international capitalists. In this context, the report argued that in the absence of political and economic reform, international aid simply provided new opportunities for ruling elites to exploit the rural poor.

Although the *Nepal in Crisis* report was not specifically concerned with issues of land degradation (its primary focus was on the impacts of road building), during the completion of the research associated with the report Piers Blaikie became aware of the devastating levels of soil erosion in the country. Sensing that the soil erosion he witnessed may be connected to the same political and economic power structures uncovered in the *Nepal in Crisis* study, Blaikie sought to develop a Marxist analysis of the processes behind soil erosion. The culmination of Blaikie's work on soil erosion came in 1985 with the publication of the influential book *Political Ecology of Soil Erosion in Developing Countries* (Blaikie, 1985).

In an account of the career of Piers Blaikie, Rigg (2006: 36) describes the significance of *Political Ecology of Soil Erosion in Developing Countries* in the following terms: 'Here was a social scientist invading the traditional turf of natural science to argue that a physical process – soil erosion – could only be understood in terms of political economy'. Given the discussions so far in this volume about the value of human perspectives on environmental change in the Anthropocene, it may seem surprising to learn that Blaikie's work was seen to be radical at the time of its publication. In many ways, Blaikie was a pioneer of the types of analyses that mix human and physical geography, and enable us to better understand the nature of the changes we are experiencing within the Anthropocene.

Box 4.3 Piers Blaikie

Born in Scotland, Piers Blaikie is a development scholar who is associated with the emergence of political ecology. Blaikie was educated at Cambridge before lecturing at Reading University and the University of East Anglia. Blaikie's early doctoral work explored patterns of rural life and livelihood in the Rjasthan and Punjab (northern India). Following research on the impacts of international aid on development practices in Nepal, Blaikie conducted an extensive analysis of the nature and origins of land degradation and soil erosion. His most recent work has focused on the spread of HIV/AIDS in Uganda.

What unites all aspects of Blaikie's extensive work is a desire to connect local issues, such as poverty, environmental degradation and disease, to broader systems of political economy. On these terms, Blaikie's work has consistently emphasized that to be successful in the long term, development policies must address the broader national and international political and economic systems that perpetuate underdevelopment and social disadvantage.

Key readings

Blaikie, P. (1985) *The Political Ecology of Soil Erosion in Developing Countries*, Longman, Harlow
Simon, D. (2007) 'Political ecology and development: Intersection, exploration and challenges arising from the work of Piers Blaikie', *Geoforum* 39: 698–707

In many ways, the early work of Blaikie on the social dimensions of soil erosion is remembered for bringing political and economic considerations into the scientific analysis of land degradation. But the notion of political ecology developed by Blaikie is as much about bringing nature into the work of social scientists as it is about bringing a human perspective into studies of nature.

4.3.2 Beyond nature and the individual

Having located the origins of political ecology within the pioneering work of Piers Blaikie, let us consider precisely what this perspective can contribute to our understanding of soil erosion. Before the emergence of political ecology, soil erosion was predominantly understood in three broad ways. First, and from the perspective of the physical sciences, soil erosion was interpreted as a natural phenomenon that, even when exacerbated by human activity, was largely driven by processes of physical environmental change. On these terms,

droughts caused by significant reductions in precipitation were seen by many to be the major cause of soil erosion. In addition to drought conditions, long periods of high winds were also seen by physical scientists to be major contributory factors within soil degradation episodes. Thomas (1993) has taken scientific understandings of the nature of soil erosion (and desertification) a step further. Thomas argues that estimates of human-induced soil erosion have been significantly overestimated, with a significant amount of what is described as desertification actually being the product of the natural fluctuation in dryland vegetation levels. Second, from a Malthusian perspective, soil erosion was associated with the pressures of overpopulation, which forced farmers and landholders to overwork their land in order to feed the expanding population (for more on Malthusian perspectives see Chapter 2). Third, and finally, soil erosion was seen as the product of bad decision-making on the part of farmers and landholders. On these terms, soil degradation was interpreted as the outcome of farmers either being

---

Box 4.4 A sandstorm in a teacup?

In a 1993 article in the *Geographical Journal*, the geographer David Thomas questioned established estimates and interpretations of desertification throughout the world. At the heart of Thomas's analysis is the belief that studies had started to confuse the natural fluctuations that exist in the spread and contraction of deserts (driven by changing rainfall levels and associated patterns of vegetation coverage) with human-induced soil erosion. According to Thomas, the conflation of these two processes resulted in vast overestimates of the rates of anthropogenic desertification. In order to ensure that better estimates of anthropogenic desertification are achieved in the future, Thomas argues that the notion of desertification should only be used in reference to dryland environments and where land degradation is not part of a natural ecological cycle. This more restricted understanding of desertification is important because it makes a distinction between human-induced soil erosion and the natural geographical spread and contraction of deserts. Thomas's work does not suggest that soil erosion is not a major global problem, but simply that it is not necessarily connected to the migration of desert fronts.

Key reading

Thomas, D.S.G. (1993) 'Sandstorm in a teacup? Understanding desertification', *Geographical Journal* 159: 318–331

unaware of sustainable agricultural practices or deliberately overexploiting soil in the pursuit of profit (see Adams, 2001: 250–255).

The political ecology perspective developed by Blaikie and his colleagues challenged each of the established understandings of soil degradation. In the context of scientific interpretations of soil erosion as largely a meteorologically driven phenomenon, Blaikie recognized that drought conditions did not always result in soil erosion. Indeed, political ecologists recognized that in more affluent societies droughts are routinely endured without significant forms of soil erosion. Clearly, while meteorological factors may trigger soil erosion events, other factors are also at play. In terms of Malthusian arguments, political ecologists claimed that advances in agriculture had, over a significant period of time, enabled farmers to produce more and more food from the land without these increases in productivity necessarily resulting in the degradation of soil. Finally, and perhaps most importantly, Blaikie and the political ecologists have been keen to avoid blaming the victims of soil erosion for land degradation processes. In this context, Blaikie's study of soil erosion in Nepal and Africa sought to understand the broader processes that were driving land degradation. On these terms Blaikie's work exposed a strong link between poverty and land degradation. It appeared that the poorest famers often only had access to the lowest quality soils. These farmers had often been removed from more fertile common lands once these lands came under the ownership of larger agro-industrial interests. In addition to land ownership issues, Blaikie's work revealed that poverty often led farmers to adopt very short time horizons in their land-use planning. The work of political ecologists revealed that when coming under pressure to feed their families poorer farmers were much less likely to employ sustainable agricultural practices (through investment in fertilizers, or leaving land fallow in order to enable it to recover from nutrient loss) (for a more detailed discussion of the links between environmental management and poverty see

Maxwell and Frankenberger, 1992). Ultimately, Blaikie's work, and the related work of other political ecologists (see Batterbury et al, 2001), has sought to understand soil erosion in the context of the political and economic processes that generate and sustain poverty. On these terms, soil erosion has as much to do with crop prices and land ownership issues as rainfall and leaching.

Since the pioneering work of Blaikie and other political ecologists during the 1970s and 1980s, political ecology has expanded to cover a range of environmental issues from a number of different theoretical perspectives (for an overview of recent work in the field see Peet et al, 2011 and Robbins, 2012). In addition to addressing ongoing issues in land degradation and soil erosion (Batterbury et al, 2001), political ecologists are also studying the impacts of nature conservation policies (Hanna et al, 2008), the consequences of resource use and extraction (Peluso and Watts, 2001), gender issues (Rocheleau, 1995) and the relation between people and their lawns (Robbins, 2007). In addition to expanding the range of environmental issues they explore, political ecologists are increasingly supplementing work in rural areas of the global south with a concern with the environmental issues of urban spaces and the ecological problems in more economically developed countries (see Keil, 2005). In the context of the expanded scope of political ecology, it is becoming an increasingly important framework within which to study the myriad of socio-environmental issues associated with the Anthropocene.

4.4 A DUST BOWL FOR THE TWENTY-FIRST CENTURY: SOIL DEGRADATION IN CHINA

It is, in many ways, tempting to think that the North American Dust Bowl (with which we started this chapter) is a thing of the past. Surely modern agriculture, replete with its inorganic fertilizers and modern scientific techniques, could never again generate soil degradation on such a scale. Yet as we make our way deeper into the twenty-first century, an episode of soil erosion that could yet match the Dust Bowl is emerging in China. The rate and scale of soil erosion in this country is now so severe that many feel it could threaten the long-term economic stability of this powerful nation.

4.4.1 The rate and scale of soil erosion in China: soil erosion at a national scale?

Between 2005 and 2007 a team of approximately 200 scientists undertook a large-scale survey in order to estimate the rate and scale of soil erosion in China. This Soil Erosion and Ecological Safety Expedition revealed some troubling trends. The survey, which looked at the soils of 7 regions and 27 provinces, revealed that soil erosion is now affecting 17 per cent of China's total land cover (Jie, 2010). This level of soil erosion sees an estimated 4.5 billion tonnes of soil removed from the land every year (*Guardian*, 2008). The problem of soil erosion in China is not, however, just restricted to one or two agricultural areas. The Soil Erosion and Ecological Safety Expedition found soil erosion in 30 per cent of all of China's counties (Jie, 2010).

Beyond the obvious ecological issues, there are significant socio-economic problems associated with China's current patterns of land degradation. Perhaps most disturbing is the prediction that current rates of soil erosion could see 100 million people in southwest China losing the land on which they currently make their livelihood (*Guardian*, 2008). According to China's own Ministry of Water Resources, current patterns of land degradation could also result in an approximately 40 per cent reduction in China's overall food production levels (Jie, 2010). Soil erosion in China is, however, also having knock-on effects for water management. As the eroded 4.5 billion tonnes of additional soil sediments enters China's rivers every year, it is leading to rising incidents of floods. This bulging sediment load is also clogging-up the country's 80,000 reservoirs, reducing their water storage capacity and increasing the threat of water scarcity (*Guardian*, 2008).

4.4.2 The causes of soil erosion

As is often the case with soil erosion, the causes of China's current soil erosion problems are multiple and complex. China's rapidly expanding economy and rising population have clearly placed pressure on the country's agricultural sector. These pressures have clearly led to the overuse and mismanagement of soil resources. Physical environmental factors also contribute to China's land degradation issues. Increasing incidents of droughts in China are clearly contributing to the country's soil management problems. There is, however, an additional environmental factor that is adding to the rate of land degradation in China. China's Ministry of Land Resources estimates that there are approximately 239,000 hectares of slope land in China (Jie, 2010). The greater the angle of slope associated with land, the more severe soil erosion processes are likely to be. On sloping land gravity greatly accelerates soil erosion rates. Given that 20 per cent of all of China's sloping land is located in arable agricultural areas, it is clear that this geomorphological condition is making soil erosion in China much worse than it would be on flatter land (Jie, 2010).

Despite these demographic and environmental forces, political ecology reminds us that it is important to place soil erosion in the context of the political and economic circumstances under which it is generated. Since the late 1970s China has been going through a period of rapid economic transformation. This transformation has seen the opening up of China's economy and the increasing application of free-market economic principles within its development strategies. These policies look likely to lead China to overtake the US as the world's largest economy. National strategies for rapid economic expansion have clearly placed pressure on the agricultural sector to produce increasing levels of food for commercial export and to feed the rapidly expanding industrial cities of

China. These pressures have been felt most keenly in China's state-run farms. These farms were formed from the properties of wealthy landlords following the communist revolution in China. Since this point, these farms have been subject to state plans for increasing food production, which often have little concern for the ecological impacts of these increases. Outside the state farm system, peasant farmers now work within communal farms. These communal farms were designed to help peasant farmers escape poverty and the yoke of unjust landowners. Many of these communal farms have been located on more marginal (often sloping land), which had been left over following the formation of state farms. In this context, communal farms have often not enabled farmers to escape poverty and have led them to adopt unsustainable land-use practices that actively contribute to land degradation.

A final contributory factor to soil erosion in China has been the large-scale forest clearances that the state has sanctioned. These forest clearances have, in part, been used to support China's expanding timber export economy (see Chapter 5). They have also enabled the freeing up of more land for agricultural food production. The end result of these deforestation processes has been the increasing exposure of often vulnerable soils to the processes of wind and water erosion and nutrient loss. Taking all of these factors into consideration, political ecology reminds us that the causes of soil erosion are rarely simple and local in origin. While soil erosion may ultimately be down to the unsustainable practices of the peasant or state farmer, or the logger, these practices are intimately connected to national land reform policies, the expanding export of MDF timber products and China's emerging role in the global economy. Addressing the problems of soil erosion in the long term would appear to require a consideration of local soil management techniques alongside global trade and national land laws.

4.4.3 Soil restoration in the Loess Plateau

By way of conclusion it is important to consider the types of remedial policies that are currently being developed in China to address soil erosion problems. Arguably the most significant remedial action to the effects of soil erosion has been developed in the Loess Plateau. The Loess Plateau is located in northwest China and, as its name suggest, is characterized by large deposits of fine-grained wind-blown soil (loess). Soil erosion has been a long-term problem on the Loess Plateau. Over a period of time a combination of heavily sloped land, fine-grained soil and unsustainable agricultural practices generated a desert-like landscape that was unable to effectively support the 50 million people who lived on the Plateau (see Plate 4.3).

During the 1990s the Chinese government and the World Bank combined forces and initiated the Loess Plateau Project. The first Loess Plateau Project saw $252 million invested in the region. This first round of investment was followed by a further investment of $239 million as part of the second Loess Plateau Project. This funding has been used to support one of the largest projects of landscape transformation seen anywhere in the world. The Loess Plateau Project has sought to transform the landscape of the Plateau from a series of steep eroded slopes to a complex network of terraces. These terraces are important because they significantly slow the rates at which soil and nutrients are transferred from the land and make agricultural practices far more productive and potentially sustainable. The project also sought to facilitate the natural regeneration of areas of the landscape by planting trees, grass and shrubs.

The World Bank claims that the Loess Plateau Project has had significant environmental and socio-economic impacts. In terms of soil erosion, they point out that sediment transfer from the area to the Yellow River has reduced by some 100 million tons a year. In socio-economic terms, the World Bank points out that since the inception of

Plate 4.3 The Loess Plateau, China
Source: Wikimedia Creative Commons, Till Niermann

the project grain yields on the Plateau have increased, as have average incomes and employment rates. The short-term gains of the Loess Plateau Project reveal the increasing power of humans, operating within the Anthropocene, to transform the environment in order to address emerging environmental problems. Political ecologists would, however, point out that without associated political and economic reform (both in the nature of land ownership and trade), there is a real danger that land degradation could re-emerge on the Plateau once project funding runs out.

More information on the Loess Plateau Project is available at:

http://www.worldbank.org/en/news/200 7/03/15/restoring-chinas-loess-plateau

This World Bank video provides information on the history of soil erosion in the Loess Plateau and how restoration efforts are attempting to reverse these problems:

http://www.youtube.com/watch? v=NQBeYffZ_SI

4.5 CONCLUSIONS

In this chapter we have explored the impacts that humans are having on the planet's soils. From the first forest clearances associated with the rise of agriculture, to the modern use of artificial fertilizers, humans have transformed soil ecologies throughout the world. The human transformation of the soil has taken four basic forms. First, through the over-utilization of soils to grow crops and rear livestock, humans have facilitated a decline in the natural fertility and nutrient content of soils. Second, through unsustainable agricultural practices, humans have facilitated soil erosion, and the associated removal of topsoil on unprecedented scales Third, through the accidental and deliberate application of chemicals humans have actively contributed to the increasing toxicity of soils. Fourth, and finally, through the accelerated application of artificial fertilizers, humans have caused a significant shift in the balance of the global nitrogen cycle, which has generated a series of harmful ecological side effects.

Throughout this chapter I have not only sought to quantify the human impacts on soil systems, but also explore the frameworks that are available to us to understand what causes soil degradation. We have thus explored scientific and Malthusian explanations of the nature of soil erosion. Drawing on the work of Piers Blaikie and the theories of political ecology, this chapter has argued that soil degradation must always be interpreted as a product of both natural (climate, slope, soil type) considerations and socio-economic processes (land ownership arrangements, poverty, global trade relations). This type of perspective helps to remind us that as societies attempt to restore soils, and establish more sustainable relationships with the land upon which we depend, we must both deploy technological solutions and instigate broader socio-economic reforms if these changes are to be long lasting and effective.

KEY READINGS

Blaikie, P. (1985) *The Political Ecology of Soil Erosion in Developing Countries*, Longman, Harlow. This volume provides a concise introduction to the work of Piers Blaikie and the formative ideas associated with political ecology.

Dent, D. et al (2007) 'Land' in UNEP (ed.) *GEO4: Global Outlook – Environment and Development*, UNEP, Nairobi: 80–114. This chapter provides a detailed statistical analysis of global patterns of soil erosion and pollution.

Smil, V. (2004) *Enriching the Earth: Fritz Haber, Carl Bosch, and the Transformation of World Food Production*, MIT Press, Cambridge. This volume provides a fascinating insight into the rise of artificial fertilizers and their impacts on human society and the environment.

Worster, D.E. (1979) *Dust Bowl: The Southern Plains in the 1930s*, Oxford University Press, New York. This book provides a detailed environmental history of the North American Dust Bowl, its causes and its consequences.

CHAPTER FIVE

Forests

Jungle capitalism and the corporate environment

5

'What happened mother, why did we run?'
'Man was in the forest'

Bambi

5.1 INTRODUCTION: THE STORY OF CHUT WUTTY

On 25 April 2012 Chut Wutty was leading journalists around a section of what he believed to be illegally logged land in the Ratanakiri region of Cambodia. Chut Wutty had been campaigning against the destruction of indigenous forestland at the hands of illegal loggers and state-sanctioned forest clearances (*Independent*, 2012). Wutty's concern for Cambodia's unspoiled woodlands led him to form the Natural Resource Protection Group. This group is based in the Cambodian capital city Phnom-Penh and lobbies government to provide better protection for its forest resources. In addition to his lobbying work, Wutty and the Natural Resource Protection Group also sought to expose government corruption and the alleged sell-off of Cambodia's national parks. On 25 April it was reported that Wutty and his party were taking photographs of illegal logging activities (BBC, 2012b). The military police officers who were present asked them to stop taking photographs and attempted to commandeer their camera's memory card. It was at this point that Wutty became involved in an altercation with a military police officer and was shot and killed (*Independent*, 2012).

While no one is claiming that Chut Wutty was deliberately targeted by the military or associated state authorities, his tragic death serves as a poignant illustration of the often-violent political and economic struggles that are waged over and within our planet's forests. On one side of this complex struggle are people such as Chut Wutty who are concerned that the large-scale commercial exploitation of forests could significantly undermine their ability to serve as precious community resources and offer important ecosystem services to the planet as a whole. On the other side are those who recognize the important commercial role of forests in providing the raw materials needed for the production of a range of everyday items including paper, nappies, furniture, rubber and even pharmaceuticals (see Dauvergne and Lister, 2011). In many ways forests are key places within which to explore the issues that confront us in the Anthropocene. Ten thousand years ago much of the Earth was covered in dense natural forests (Dauvergne and Lister, 2011). In many respects the gradual clearance of these primordial forests is the most obvious visible sign of the rise of the human geological era (Ruddiman, 2005). This chapter charts the evolving relationships between humans and forests and considers the historical and contemporary forces that have shaped these relations. Particular attention is given, in this context, to the increasingly significant role of multinational timber and paper corporations and global commodity chains in driving forward forest clearances of different kinds. This chapter also considers the role of governments and NGOs in attempting to conserve forest resources and in developing more sustainable forest management practices.

This chapter begins with an account of the impacts of human economic development on global forest resources. The section that follows outlines the role of corporations and the global market place in shaping human relations with forests. The final two sections of this chapter introduce case study examples. The first considers the operations of the United Fruit Company in the jungles of Central America. The second case study explores the operation of 'big box' timber retailers and their role in coordinating the global timber commodity chain.

5.2 TRANSFORMING FORESTS: REFLECTIONS ON THE LONG ANTHROPOCENE

As I sit writing this chapter in the hills of west Wales, the view through my window appears to be quintessentially natural. The rolling hills, sheep pastures and small clutches of trees appear to be timeless and eternal. In fact nothing could be farther from the truth. If we return to the pre-agricultural period of 6000 to 10,000 years ago, much of the area I can see through my window would have been covered in woodlands (see Kaplan et al, 2009). There is still much scientific debate concerning the nature of the woodland that existed at this point in time, with some suggesting a fairly continuous, closed-canopy style woodland, while others claim that the early Holocene landscape was a mixed landscape of 'pasture woodland' (Coelho, 2009).[1] What is more, the scattered pieces of wood-lands I can see through my windowpane today are not even remnants of this older forest. They are either modern industrial pine plantations, or second growth woodlands that have gradually repopulated this area. So what has been happening to this arboreal landscape over the last 10,000 years?

Things really began to change in the nature of forest coverage in Europe in 4000 BCE. It is at this point that human societies started to shift their lifestyles from the hunting and gathering of food to more sedentary agricultural ways of life (see Chapter 4 for a more detailed discussion of the agricultural revolution). The growing of crops and rearing of domestic animals had a series of profound impacts on human civilization. As we will see in Chapter 6, this agricultural shift enabled the birth of the first large-scale cities. But this transformation in the way in which humans produced food also had significant environmental consequences. In order to develop effective agricultural systems, early human communities had to clear away the woodlands that would have obscured the light needed to grow crops and made animal husbandry difficult to achieve. Over time forest clearance practices were also driven by the need for fuel and the demand for wood in home and ship construction industries. The processes that led to the clearance of woodlands some 6000 years ago mean that today more than half of the primordial forest that covered the Earth has now been felled (see Dauvergne and Lister, 2011: 2). But the rate of forest clearance is accelerating. Since 1950 the deforestation of original woodlands has been equivalent to the rates of loss that had occurred over the previous 6000 years (Dauvergne and Lister, 2011: 2).

Things are, however, more complicated and ecologically problematic than these aggregate figures suggest. If we consider the geography of deforestation in recent years, we notice significant changes in its distribution. In many Western European nations there is a now a drive towards the preservation of ancient woodlands and refor-estation. In other areas of the world, deforestation is accelerating. In Russia, for example, the boreal forest (forests found in cold sub-Arctic climates) is being heavily exploited as part the country's resource-based strategy for economic development. In emerging economies, such as Brazil and China, deforestation has provided an important basis for economic expansion as these countries seek to exploit the lucrative global timber market (Dauvergne and Lister, 2011: 3). In tropical areas, rainforests are being felled at alarming rates in order to feed the global timber industry, but also to make way for more lucrative agricultural practices (such as ranching) to occupy the land where forests once stood. Estimates from the FAO indicate that it is countries in the global south, and, in particular, states that contain belts of tropical rainforests that are now being deforested at the fastest rates (FAO, 2007). In South America, for example, deforestation rates are running at 4.3 million hectares a year (FAO, 2007). On the African continent some 4 million hectares of forests are being lost annually, while the state of Indonesia is, on its own, losing an estimated 1.9 million hectares of forestland a year (FAO, 2007).

Forest clearances not only affect the local communities of people and species that depend upon them; they also have significant implications for the global environment. In his theory of the 'early' or 'long Anthropocene', William Ruddiman has argued that it was actually the early clearances of forests some 6000 years ago that marked the

beginning of the Anthropocene and humankind's ability to affect global environmental systems (Ruddiman, 2005). More worrying, however, is the impact that the current clearance of tropical rainforests is having on the global climate. While estimates of the contribution of forests to the global carbon balance vary greatly, it is believed that tropical rainforests absorb over 1 billion tonnes of carbon from the atmosphere on an annual basis (Dauvergne and Lister, 2011: 2). Furthermore, the act of deforestation itself is now believed to contribute to the release of around one fifth of humanly produced carbon dioxide (Dauvergne and Lister, 2011). What these estimates reveal is that it is unlikely that attempts at reforestation in more temperate forest areas are going to offset the huge climatic impacts of deforestation in tropical zones.

Recognizing the significant problems that are associated with deforestation in less economically developed countries is not to suggest that the problems of deforestation in such places are only the responsibility of developing nations. Figures 5.1 and 5.2 reveal changing patterns in the trade

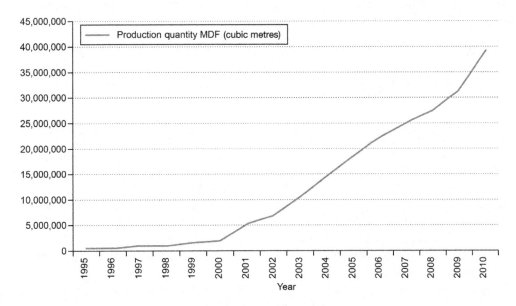

Figure 5.1 Total MDF production in China (1995–2010)
Source: FAOSTAT

Box 5.1 *Plows, Plagues, and Petroleum*: reflections on the long Anthropocene

The human clearance of forests may represent an important historical marker in the history of the Anthropocene. The palaeoclimatologist William Ruddiman has argued that it is not the rise of modern industry that marks the beginning of the human geological era, but the emergence of widespread deforestation. In his book *Plows, Plagues, and Petroleum*, Ruddiman (2005) develops what has been termed an 'early' or 'long Anthropocene' hypothesis. According to Ruddiman there is a distinct pattern to the changing concentrations of greenhouse gases in the Earth's atmosphere since the last ice age (some 11,000 years ago). While in previous ice ages levels of carbon dioxide and methane reached a climax just after the ice age and then declined, in the Holocene something different has happened. While levels of greenhouse gases did fall for around 3000 years, they then started to rise and have continued to do so ever since. Clearly these recorded rises in atmospheric carbon dioxide and methane levels cannot all be attributed to modern industrialization, which really only began some 200 to 300 years ago. According to Ruddiman's long Anthropocene hypothesis, the reason that the current inter-glacial period has seen a long-term trend in rising levels of greenhouse gases can only be attributed to the rise of agricultural society. The large-scale forest clearances associated with the emergence of agriculture resulted in a significant reduction in the planet's natural capacity to absorb and store carbon dioxide. In addition to this, Ruddiman connects the emergence of rice cultivation some 5000 years ago with rapid increases in global levels of methane (the wet and warm conditions associated with rice paddies are ideal for the process of methanogenesis).

Ruddiman's hypothesis has been challenged by some (see Peterson, 2008). Counter arguments claim that too few people were involved in agriculture 8000 years ago to account for the increases in greenhouses gases observed. It is also claimed that there is no real norm in the pattern of greenhouses gas levels in past glacial and inter-glacial periods, so it is problematic to claim that our current situation is necessarily unique (Peterson, 2008).

Key readings

Ruddiman, W.F. (2001) *Earth's Climate: Past and Future*, W.H. Freeman and Co, NY
Ruddiman, W.F. (2005) *Plows, Plagues, and Petroleum*, Princeton University Press, Princeton

of MDF products in China and the UK. MDF, or medium-density fibreboard, is an engineered wood product that is comprised of hardwood and softwood residuals. MDF is commonly used to produce household furniture and shelving. What Figures 5.1 and 5.2 reveal is that over the last 17 years China has been rapidly increasing its production of MDF, while the UK has seen a significant increase in the levels of MDF that it imports into its domestic market. What these figures ultimately reveal is an emerging global market in MDF products. This market is based upon rapid deforestation in countries such as China, and the associated relocation of multi-national wood manufacturing facilities to such low-cost locations. This market is also, however, based upon a growing consumer market for such wood products within more economically developed countries, such as the UK. It is in this context that understanding the forces now driving deforestation requires an appreciation of the nature of globalization and corporate environmental relations.

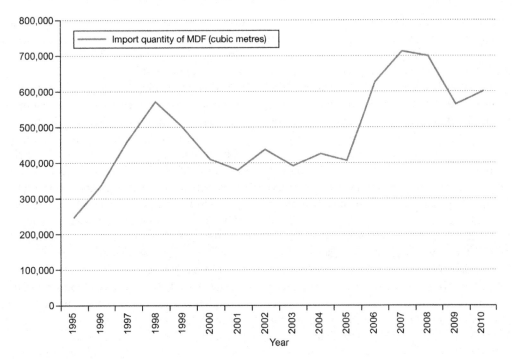

Figure 5.2 Quantities of imported MDF to the UK (1995–2010)
Source: FAOSTAT

5.3 GLOBALIZING THE FOREST AND MULTINATIONAL FOREST CORPORATIONS

As the previous section outlined, the main environmental concerns related to contemporary forms of deforestation are not just about the rate of forest clearances but the locations where these clearances are taking place. Unlike energy resources such as coal, oil and gas (see Chapter 2), forests are a renewable resource, and as such will always be able to be replenished to some degree (see Dauvergne and Lister, 2011: 1). The main problem associated with contemporary deforestation is that it tends to be concentrated in environmentally sensitive locations. This concentration process means that forest clearances can seriously erode the social and ecological services that woodlands provide at both local and global levels. In order to understand the contemporary geography of deforestation, we need to grasp two interconnected things: *globalization* and the *multinational corporation.*

5.3.1 Unpacking globalization

While the notion of globalization has been defined and interpreted by different people in very different ways (see Held et al, 1999), it is generally associated with a series of key features. At a relatively simple *economic* level, globalization is associated with increasing levels of integration and connection between economic activities throughout the world. These economic connections can relate to production processes, where we now see the different parts that comprise consumer products such as cars and computers being made in different regions of the world before they are integrated into their final form. These connections can also be seen in relation to the market place for goods, where increasing amounts of goods are being purchased and used in countries that are a long way from the places in which they were grown or made. A final, if often obscured, aspect of economic globalization is the ease and speed with which financial transactions can now travel around the planet. The financial dimension of globalization is critical as it enables large sums of money to be drawn together and invested into the economic ventures that drive the aforementioned processes of production and consumption (see Harvey, 2006). But globalization is not merely a set of economic processes; it is also inherently political (see Stiglitz, 2002). A series of international organizations have been formed in order to oversee and regulate the economic globalization process. These organizations include the World Trade Organization (which regulates international trade), the International Monetary Fund and the World Bank (which provide financial support and investment for countries in order for them participate in the global market place).

So it is possible to identify the presence of globalization on the basis of globalized forms of economic activity and related global political institutions. But these identifying markers of globalization tell us very little about why globalization exists. In many respects, globalization is the outcome of an international community that is dedicated to market-oriented forms of society. Market-oriented societies are based on the assumption that the free market place is simultaneously the most efficient way of producing and circulating goods and of preserving personal freedom (see Hayek, 1960; Friedman, 1982). In the first instance, those who advocate market-oriented forms of social organization claim that it is the most effective way to ensure that people get what they want, in the quantities that they want it, and at the most reasonable price. The reason for this, it is claimed, is that in a free market place entrepreneurs are given the necessary freedoms to follow their profit-making instincts by making products that consumers appear to prefer. Furthermore, because they are competing with

other producers within the market place, such entrepreneurs have a clear incentive to search for innovative ways to produce better goods at lower prices. In the second instance, the market place is seen as a place within which personal freedom can be preserved. As long as market exchange is voluntary (which, as we will see later, is often not the case), it can enable effective social coordination without requiring oppressive forms of coercion (Friedman, 1982: 13). In a market-oriented society, consumers can choose which goods they wish to buy from who, and who they wish to sell their labour to in the workplace. If they are unhappy with the goods they buy, or the conditions of their work, they can choose to shop and work elsewhere (or at least it is argued so).

The expanded economic freedoms associated with globalization support these market-based visions of society in important ways. First, globalization can be seen to increase the field of competition of those making and supply goods and services to consumers (although this is not always the case, see Box 5.2 on Joseph Stiglitz below). Second, globalization expands the market place within which successful entrepreneurs can sell their products and services. Third, globalization makes it easier for investment to flow to successful economic enterprises in diverse geographical locations, which makes it easier for enterprises to grow.

5.3.2 The multinational corporation

If globalization embodies an increasingly significant way of organizing economic life in the twenty-first century, there is one type of organization with which it is particularly synonymous: the multinational corporation. Multinational corporations (hereafter MNCs) come in a range of forms and sizes, but they have one thing in common. What connects MNCs is their ability to organize their economic activity at an international scale, and to be able to move these activities between a range of different countries. The geographical flexibility of MNCs provides them with significant economic

> ## Box 5.2 Joseph Stiglitz: a critique of globalization
>
> Joseph Stiglitz is a Nobel Prize-winning economist. He served on the US Council of Economic Advisors under the then President Bill Clinton. In 1997 he moved to the World Bank where he served as chief economist and senior vice president until 2000. In his best selling 2002 book, *Globalization and its Discontents*, Stiglitz uses his unique insights into the operations of the global economy to develop a critique of globalization. While Stiglitz recognizes the potential of globalization to enhance the lives of people throughout the world, he argues that the form it is presently taking is having a detrimental impact on the freedoms and lifestyles of many throughout the world. At one level, Stiglitz notes how the trade agreements that form the basis for globalization tend to be forged primarily in the context of the special interests of economically powerful nations and corporations. This process tends to perpetuate unfairness in global trade relations and can be a real barrier to development in many less economically developed countries. At another level, Stiglitz notices how financial aid packages offered by organizations such as the International Monetary Fund to countries that are in financial turmoil tend to be based on naïve and out-dated economic assumptions. In this context, countries seeking financial aid are expected to go through a process of structural adjustment in and through which their economies are made 'globalization ready'. The process of structural adjustment involves the deregulation of national economies, so that they become more market-oriented and efficient, and the opening up of these economies to the 'opportunities' of global trade and investment. According to Stiglitz, these market-based ideologies tend to overplay the potential of global markets to improve economic efficiency and social opportunities. Stiglitz also argues that such market-based ideologies tend to down play the positive role that state intervention in the market place can have.
>
> ### Key readings
>
> Harvey, D. (2005) *A Brief History of Neoliberalism*, Oxford University Press, Oxford
> Peck, J. (2010) *Constructions of Neoliberal Reason*, Oxford University Press, Oxford
> Stiglitz, J.E. (2002) *Globalization and its Discontents*, Penguin Books, London

advantages. At one level, MNCs, like the fast food outlet McDonald's, operate in a number of different countries in order to expand the consumer population that can purchase its fast food products. Other MNCs, like the energy giant ExxonMobil, operate globally because the source of materials upon which this company relies (in this case gas and oil) are concentrated in certain countries and regions throughout the world. MNCs such as the clothes manufacturer GAP, have, by contrast, sought to globalize their production of apparel as a way to exploit the lower labour costs that can be found throughout different parts of the world.

While MNCs are diverse, their significance as agents within the Anthropocene should not be underestimated. By 1995, it was estimated by the United Nations Conference on Trade and Development that 40,000 MNCs controlled two thirds of all global trade in goods and services. Perhaps even more telling than this statistic, however, is the rising power of corporations compared to nation states (see Chapter 7). Of the one hundred most wealthy global organizations (which include nation states), 66 are now corporations (Gray, 1998). In addition to this, by the late 1990s, only 57 nation states had budgets that would have enabled them to be on the Forbes 500 list of the world's most wealthy corporations (Gray, 1998). It is in the context of such power – even compared to the nation states that are tasked to regulate them – that MNCs have been subject to sustained critique. Some argue that the size and power of MNCs now means that they are able to create virtual price monopolies for the goods and services they provide, and thus ultimately reduce competition and choice for consumers. In his analysis of Walmart, for example, Joseph Stiglitz, recognizes how large food retailers can drive out local food supplies by undercutting their prices (2002: 68). In the long run, however, once competition is removed in the local area it is all too easy for supermarket giants to increases their prices (Stiglitz, 2002). Others claim that corporations have become so powerful that they can influence the decisions that are made by elected governments and international organizations, such as the International Monetary Fund and World Bank, in order to further their own economic interests (Monbiot, 2000). In specifically environmental terms, critics have also argued that the geographical freedoms of MNCs have made it easy for them to exploit locations where environmental standards are lower (particularly in less economically developed countries) and expensive anti-pollution measures and waste treatment can be avoided (Harvey, 1996: 366–369).

The Corporation

This documentary by Mark Achbar, Jennifer Abbot and Joel Bakan provides an excellent introduction to the history, legal form and economic rationale of the corporation. It famously concludes with a psychological profile of the corporate form and suggests that it displays many of the features of a psychopath.

5.3.3 Globalizing the chainsaw

It is now time to return to our discussions of forestry and the ways in which the processes of globalization and the practices of MNCs shape human–forest relations. At one level globalization has now made it much easier for large timber and paper MNCs to exploit the rich timber resources within the tropics. While some 80 per cent of global forests are publically owned, many less economically developed countries are keen to grant logging permits to MNCs as a basis for securing (often short-term) investment and employment opportunities within their countries (Dauvergne and Lister, 2011: 9). Dauvergne and Lister (2011) argue that MNCs often have very little incentive to sustainably harvest forest in less economically developed countries, and will often exploit woodland resources as quickly as they can before moving on to other areas (Dauvergne and Lister, 2011:11). Such practices can have a devastating impact on the ecological integrity of forests and the communities that depend upon them. But the model of the foot-loose multinational logging corporation as the vertically integrated framework within which deforestation is coordinated is being challenged by other modes of economic organization. Dauvergne and Lister (2011) chart the rise of more loosely connected timber commodity chains, which comprise of local small-scale logging organizations and large-scale wood processing outlets. Large corporate retailers (such as IKEA and B&Q) often head-up these economic alliances and are able to use competition between small-scale timber producers in order to drive market prices down and increase their own competitive advantage.

The impact of globalization on forest resources stretches beyond the actions of the international timber industry. In many parts of the world, the clearance of tropical woodlands has been driven by agricultural interests who are keen to exploit the agricultural lands that are found beneath the canopies of rainforests (see Chapman, 2007). As with multinational timber operations, these agricultural practices are often coordinated by global food corporations who are eager to exploit tropical climates in order to produce high-profit yielding commodities such as beef and bananas.

The globalization of forest resources has resulted in more than officially sanctioned deforestation. In many instances, so-called illegal logging (often in protected forest areas) is informally supported by governmental authorities, which protect timber corporations from legal prosecution. At another level, however, small-scale illegal logging is on the rise (INTERPOL/World Bank, undated). Recent estimates by the World Bank suggest that illegal logging activities throughout the world result in economic losses of some $10 billion every year, plus lost tax and royalty payments of $5 billion per annum (World Bank, 2006). In many states illegal logging also constitutes a significant percentage of all logging activities (see Figure 5.4). The extent of current illegal logging activities is clearly the product of the great profits that can be made within the global timber markets. The processes of globalization do, however, also make it much easier to traffic illegal timber throughout the world. As illegal timber is transported throughout the world its association with illegal points of origin can be easily obscured.

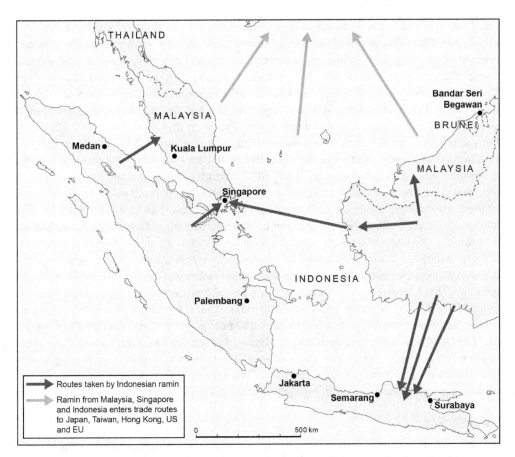

Figure 5.3 Routes of illegal timber trade in Southeast Asia
Source: INTERPOL

When we consider the varied processes associated with the globalization of forest resources together – including MNC activity, global timber processing and supply chains, the operations of the agro-industrial sector and the clandestine business of illegal loggers – what we find is that the conditions of forests are now intimately tied to the ebb and flow of global capitalism (see Prudham, 2005). To these ends, the rate of forest clearance (both legal and illegal) and the types of trees that are being felled are connected to the fluctuating prices of different timber products on the global market. Ultimately, external forces and distant decision-makers, who have little sense of the broader biological integrity and social value of woodlands, now determine what goes on in different forests throughout the world.

Box 5.3 INTERPOL and illegal logging: reflections on the Chainsaw Project

The Chainsaw Project is a partnership between INTERPOL and the World Bank that was initiated in 2007. The idea behind the project was to connect environmentally oriented research on the long-term ecological impacts of illegal logging to discussions of international criminal justice. INTERPOL is the only international police organization, and as part of the Chainsaw Project it sought to use its expertise in studying international forms of crime to understand the nature and extent of the illegal timber industry. The Chainsaw Project report sought to move beyond understandings of illegal logging that interpreted it as a series of fairly sporadic, local activities. The Chainsaw Project report thus defines illegal logging (or trafficking), '[a]s a succession of criminal activities undertaken at an international level by a network of organized criminals' (INTERPOL/World Bank, undated: 3). This shift in emphasis was based upon the realization that rising timber demand and the globalization of the timber supply chain are now providing new opportunities for illegal logging. This shift is also based upon the fact that the profits accrued from illegal logging are being accumulated by international criminal organizations and strategically reinvested in order to expand their illegal logging capability. It is precisely in these contexts that illegal logging is now a significant contributory factor within global deforestation. Indeed, INTERPOL now estimates that a forested area equivalent to the size of Austria is lost to illegal logging every year (INTERPOL/World Bank, undated). This means that the amount of illegally produced timber represents somewhere between 20 and 50 per cent of the total global timber market (INTERPOL/World Bank, undated). On the basis of these figures, and the fact that illegal logging tends to occur at an international level, the Chainsaw Project report stresses the importance of establishing an environmental crimes sub-directorate within INTERPOL in order to provide effective support for the monitoring and prosecution of criminal activities.

Key reading

INTERPOL/World Bank (undated) *Chainsaw Project: An International Perspective On Law Enforcement Illegal Logging*, INTERPOL/World Bank, Rome

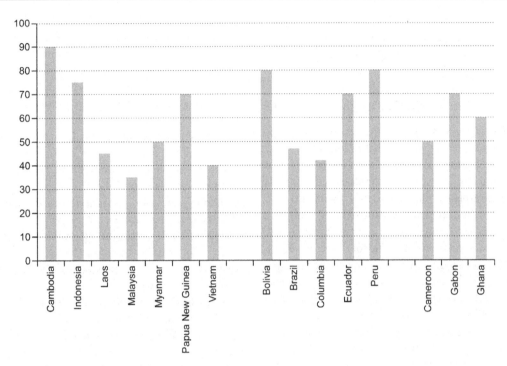

Figure 5.4 Estimates of legal logging as a percentage of total logging activity

Note: Where only estimate ranges were available, the higher estimate has been included here.
Source: INTERPOL/World Bank, undated: 6

Plate 5.1 Deforestation of the Atlantic Forest, Rio de Janeiro, Brazil
Source: Wikimedia Creative Commons

5.4 JUNGLE CAPITALISM: THE CASE OF THE UNITED FRUIT COMPANY

5.4.1 The United Fruit Company and the early drive to globalization

The first case study we consider concerning the connections between globalization, MNCs and deforestation is provided by the United Fruit Company. The choice of this case study requires some explanation. The need for explanation stems from the fact that the United Fruit Company (hereafter UFC) is not a multinational timber corporation or major retailer of timber products. Notwithstanding this point, this section focuses on UFC for three important reasons: 1) because the activities of UFC are an early example of the globalizing activities of a truly MNC; 2) because of the significant impact that the activities of UFC had on the tropical rainforests of Central America; and 3) because UFC provides a salutary lesson in what the unchecked power of MNCs can lead to.

The UFC was officially established on the 30 March 1899 (Chapman, 2007). By this point, however, it already owned land in Costa Rica, Panama, Colombia, Cuba, Jamaica and the Dominican Republic, which collectively constituted some 50,000 acres (Chapman, 2007: 49). In essence, UFC was an amalgamation of the America businessman Minor Cooper Keith's tropical fruit interests in Central America and the Boston Fruit Company. Over the course of the twentieth century, UFC would become one of the dominant players in the international fruit trade and almost develop a total monopoly on the production and transportation of bananas. The only reason that you may not have heard of this company today is because it was renamed Chiquita Brands International in 1984. The next time you buy a Chiquita banana (with its famous blue and yellow label, replete with an image of a gyrating woman wearing a hat full of fruit), you are forming a direct connection with the story we are about to hear.

Following its formation in 1899, UFC endeavoured to increase its fruit production and supply capacity. It did this through investment in its shipping fleet (its so-called Great White Fleet), which would bring its bananas to North America, and through the acquisition of more land in tropical states where banana production was most productive. Building on Minor Cooper Keith's existing political connections in Central America, UFC acquired land and won valuable tax breaks for its operations. Over time, UFC expanded its corporate interests and starting to buy up radio broadcasting companies, postal services and established sugar and palm oil plantations (see Figure 5.5). During the middle decades of the twentieth century, UFC was one of the largest employers in the whole of Central America and acquired significant political influence in all the countries within which it operated. It was in the context of such circumstances that UFC started to exploit its global reach and corporate power in ways that were socially and ecologically damaging.

In his thoroughly engaging account of the activities of UFC, *Jungle Capitalists: A Story of Globalization, Greed and Revolution*, Peter Chapman (2007: 7) describes the corporation's economic philosophy in the following terms:

> United Fruit set the template for capitalism, the first of the modern multinationals . . . There were older companies than United Fruit . . . but they had been stay at home types . . . It had set up its own enclave in Central America, a network of far-flung plantations and company towns that acted as an experimental laboratory for capitalism.

Chapman recounts how UFC was able to use its power and influence to ensure that its economic activities in Central America were free from unwanted governmental regulations and intervention. Chapman describes the extraordinary lengths that UFC would go to in order to preserve their operational freedoms through the case of Guatemala in the 1950s. Following the election

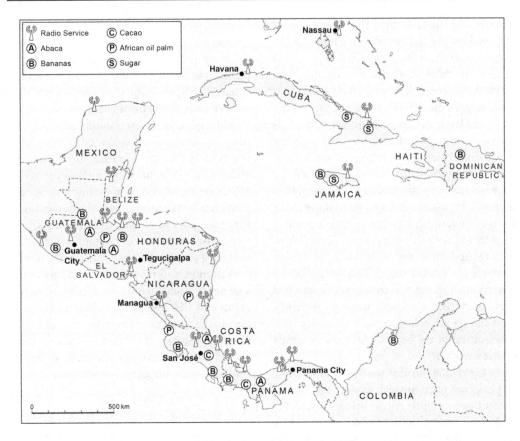

Figure 5.5 United Fruit Company activities in Central America in 1951
Source: Adapted from *Life Magazine*, 1951

of President Jacobo Árbenz in 1951, UFC became concerned about the incoming president's proposed reforms. At the heart of these reforms was a plan to redistribute uncultivated lands, which were in the possession of large MNCs like UFC, to individual families. Guatemala's banana plantations accounted for a quarter of all of UFC landholdings in South America (Striffler and Moberg, 2003). UFC consequently utilized its connections with the CIA in order to orchestrate a coup d'état in Guatemala. In 1954 Árbenz was ousted from power and his government replaced with a military junta that was much more sympathetic to the needs of UFC (see Chapman, 2007: 127–141).

5.4.2 Banana republics and socio-environmental injustice

In many ways the case of UFC's operations in Guatemala reflects a kind of extreme expression of the levels of political influence that global corporations can have over the states within which they operate. In order to be free from the social and environmental restrictions that may be placed upon them within their countries of origin, MNCs seek out weaker, less-powerful states, and exploit the favourable economic conditions they find there. Indeed, the very phrase 'banana republic' – a pejorative term used to describe a weak and possibly corrupt state that becomes the servant of narrow corporate interests – was developed to describe the impacts that the operations of corporations like UFC were having in places like Central America. In this context, it is important to consider in more detail the impacts that such political and economic activities had on the forest ecosystems in which UFC operated.

At one level, UFC was able to use its unchallenged power to gradually acquire more and more of Central America's first-growth rainforests. These forests were either cleared to make way for large banana or sugar plantations, or simply left unused. UFC would often take ownership of large tracts of land, which it had no intention of using, in order to prevent its competitors being able to acquire productive land from which they could compete for market share. This land-banking process often meant that communities who had previously relied on forests to secure precious food and resources had to move elsewhere (this problem was at the heart of Árbenz's proposed land reforms).

In the plantations that were formed by UFC, emphasis was placed on high levels of agricultural production with little regard being given to the ecological sustainability of the practices. The emphasis that UFC placed on high-yielding agricultural output resulted in it concentrating most of its plantation activities on the growing of one particular type of banana: the Gros Michel, or Big Mike (Chapman, 2007). The Big Mike is actually the larger type of banana you will come across when perusing the fruit shelves of your local green grocer. The problem was that in concentrating its efforts on the growth of the Big Mike, UFC created a monoculture plantation system throughout Central America that was prone to disease. It was in this context that Panama disease was able to sweep through UFC plantations and decimate its banana harvest. Given its relatively unchallenged power in Central America, and its access to large swathes of forested land, UFC's solution to the problem was simple: it moved its activities around. Chapman (2007: 105) thus describes how 'United Fruit's response when the disease had taken hold was to move on to new

land, to a new country if need be, and to carve out another part of Central America's infinite jungle'. As more jungle was replaced by single-crop plantations, the diverse sets of plant and animal species that depended on these forests found their survival increasingly threatened.

Although UFC is not a multinational timber company, its operations in Central America reveal the impacts that the processes associated with globalization can have on forest ecosystems. MNCs are able to use their economic power and political influence in order to avoid the regulations that may exist in certain countries to protect forest ecosystems. They often employ ecologically damaging practices that can result in the need to convert more and more forests into agricultural land just to maintain food production levels. The example of UFC also indicates how, in the age of globalization, the fate of forests becomes dislocated from the places in which they are located, and is instead connected to economic decisions that are made in distant corporate boardrooms.

5.5 BIG BOX RETAIL AND THE GLOBAL TIMBER SUPPLY CHAIN

5.5.1 IKEA arrives!

I remember vividly when IKEA first opened a branch in my hometown. I must have been about 14 at the time, and had a weekly paper round. As part of my paper delivery duties I would also distribute fliers advertising local businesses. One week I was given a large and eye-catching brochure to slip into the papers I was delivering. The brochure was a distinctive blue and yellow colour and was emblazoned with, what seemed to me at the time at least, the exotic sounding name 'IKEA'. I remember as I delivered those brochures feeling a sense of excitement: it seemed like something of great corporate importance, which held the promise of new exciting lifestyles, had arrived. My sense of IKEA's significance was not diminished when I caught my first sight of the store as I passed

Plate 5.2 The first IKEA store, Älmhult, Sweden
Source: Wikimedia Creative Commons, Christian Koehn

it on the motorway. Surrounded by an obedient legion of yellow and blue flags, the box-like store dominated the skyline and made the council houses that it stood next to look minuscule. Today, I am somewhat less impressed with my home-town's IKEA, but I am in no doubt of its commercial significance.

IKEA was formed by Ingvar Kamprad in Sweden in 1943. Since its inception during World War II, IKEA has now grown to be the largest furniture retailer and the biggest retailer of wood-based products in the world (Dauvergne and Lister, 2011: 34). It offers over 9000 timber-based products to its customers and consumes somewhere in the region of 7 million cubic metres of wood per year (Dauvergne and Lister, 2011: 35). With its 301 stores in 37 different countries, IKEA is the example par excellence of the globalization of timber retail. But IKEA is not alone when it comes to the growing impact of big box retail on

global timber transactions. Walmart, Home Depot, Kingfisher (B&Q), Staples and Lowe's are all significant players in the global trade of timber. In this section I want to consider the impacts that big box retailers are having on the trans-formation of the world's forests. In particular, we see that the increasingly complex chains of global timber supply are leading large retailers to (often unwittingly) support the illegal felling of ecologically important forest resources.

5.5.2 Economies of scale in the contemporary timber market

At the heart of the economic model of large retail companies such as IKEA, Home Depot and B&Q is the principle of the economies of scale. The idea of economy of scale is based on the fact that the greater the quantity of economic products you produce and sell, the easier it is to keep prices down and remain competitive in the market place. And this is precisely what large timber retailers have been doing in recent years: they have flooded home improvement and furniture markets with products that are so cheap that it is impossible for smaller, local retailers to compete (Dauvergne and Lister, 2011: 39). But in order to understand how this economic model actually operates we need to return to our discussions of globalization.

In their fascinating book *Timber*, Dauvergne and Lister (2011) describe the ways in which the commodity chains associated with wood products have been transformed since the 1990s. The traditional timber commodity chain involved the movement of wood fibre from logging operations to so-called primary manufacturers (including sawmills) and on to secondary manufacturers, who make higher-grade products such as furniture or kitchen fittings (Dauvergne and Lister, 2011: 41–42). The timber products would then move to wholesalers, who would in turn supply retailers. Big box retailers have been able to exploit the economic advantages associated with globalization in order to utterly transform this system. The processes of globalization have enabled an expansion in the number of suppliers of timber products, who are all suddenly able to plug into the system of global trade without being stuck in rigid and hierarchical commodity chains. Large timber retailers have sought to exploit the large number of suppliers in order to pursue the lowest cost sources of wood. These retailers have also attempted to reduce the price of the timber they receive by removing many of the links in the old wood supply chain (including wholesalers who would have taken their cut of the sell-on profits). In doing so, large retailers have also been able to connect themselves more directly with their primary suppliers. Having reorganized supply chains in their favour, Dauvergne and Lister (2011) describe how big retailers then 'squeeze' manufactures into reducing their prices further. This drive to squeeze prices is achieved in two ways: 1) large retailers will place exceptionally large orders with manufacturers (often in the tens of millions of dollars), and expect a discount to reflect this commitment; and 2) in these days of flexible supply chains, modern retailers can also threaten to change supplier if a manufacturer does not match their asking price (Dauvergne and Lister, 2011: 48–49).

In essence globalization has enabled large timber retailers to exploit economies of scale at both ends of the supply chain. At one level, they can demand low prices from the suppliers who they have increasing influence over through their ability to place lucrative orders. At the other end of the supply chain, through their presence on the high street (or more often now out-of-town retail locations) throughout the world, big box retailers can saturate local markets with their products and secure an ever-bigger share of the consumer market. This model has been described by some as a system of *low costs–high control* (Dauvergne and Lister, 2011). The question remains, however, as to what the environmental consequences of this new model of timber production and supply are.

5.5.3 The environmental consequences of big box retail

The environmental consequences associated with contemporary timber commodity chains fall into two categories. First is the increasing aggregate demand that is being placed on forest resources. A key part of the logic of big retailing is that low costs will drive up effective demand for timber products, with more people being able to afford them. Added to this are clever lifestyle marketing strategies that are developed by retailers such as IKEA. As you walk around their big box units you find full re-creations of not just kitchens, bathrooms, lounges and bedrooms, but of whole lifestyles. This IKEA chic can range from loft living in New York City to an English country kitchen. At its heart, this form of marketing seeks to develop a form of consumer demand that is not based on need but on aspirations for new ways of living (see Klein, 2010). Through the combination of low prices and marketing, big box retailers are clearly contributing to the escalating demands that are being placed on the world's forests to supply timber to the market place.

The second set of environmental concerns relates both to where big box retailers are sourcing their timber and the conditions under which it is being produced. In their reflections on contemporary timber sourcing strategies, Dauvergne and Lister (2011: 49) observe that:

> Manufacturers and timber producers find creative – and sometimes illegal – ways to lower prices for big buyers . . . this has meant lowering employee wages and health-and-safety measures, purchasing more illegal timber, and adopting destructive forest practices.

In a report on the timber sourcing strategies of Walmart in China and Russia, the Environmental Investigation Agency (EIA, 2007) noted a lack of concern within the corporation with precisely where the wood that went in their products was coming from. To use one specific example, the Environmental Investigation Agency noted how the baby cribs that Walmart sold were made from wood that was sourced in areas of Russia that had very high rates of illegal logging, which is often carried out during the tiger breeding season. On this basis, the Environmental Investigation Agency concluded that it was likely that Walmart products contained illegally sourced timber whose harvesting was having a detrimental effect on forest ecosystems and associated levels of biodiversity. While Walmart may itself be oblivious to the presence of illegal timber within its products, it is clear that its economic practices and associated global supply chains create the incentives and opportunities that lead to the exploitation of protected woodlands.

Of course, Walmart is not alone in its connections to illegal sources of timber. Carrefour has been criticized for selling wood furniture from protected forests in Vietnam and Malaysia, while Home Depot has sold mahogany products that

> Find out more about the work of the Environmental Investigation Agency at its forest loss site:
>
> http://www.eia-international.org/ our-work/ecosystems-and-biodiversity/ forest-loss

were sourced in the Brazilian Amazon. It is, however, encouraging to note that following media coverage of such controversial sourcing practices that many big retailers are trying to take greater responsibility for precisely where their wood is coming from. In the case of IKEA, for example, the company now has an environmental programme that, among others things, commits it to sourcing its timber from sustainable sources. In order to support such corporate greening activities, there are now official forest certification processes. These initiatives are often led by environmental NGOs such as the Forest Stewardship Council, and provide an official labelling system for timber products that have been produced in socially and environmentally just ways (see Eden and Bear, 2010).

5.6 CONCLUSIONS

In his book *In Amazonia* the anthropologist Hugh Raffles (2002) describes nineteenth-century attempts to cut artificial canals that would extend the Amazonia waterway and enable a renewed wave of settlement and exploitation in Amazonia. Raffles describes the ultimately lost battle of these early canal engineers and their attempts to tame the vastness of Amazonian nature. In Raffles's account we find humans confronting nature on an epic scale. This is a form of nature that could easily swallow up and reoccupy ground works that humans had spent years constructing. I mention Raffles's account of Amazonia here because it reminds us of the enduring capacity of forests to push back against the best efforts of humankind. The point is that within our accounts of the

Anthropocene there is a tendency to depict nature as a helpless victim when confronted with the all-powerful desires and technologies of humans. In a related sense, Alan Weisman's recent book *The World Without Us* reminds us that should humans disappear from the planet today, it would probably only take around 400 years for nature and forests to reoccupy vast areas of the planet (Weisman, 2007). Notwithstanding these observations, within this chapter we charted a distinctive shift in the ability of human beings to transform the global forest balance.

In this chapter we described the processes in and through which human activity has led to the gradual, but now accelerating, loss of forest cover throughout the world. This process began some 6000 years ago with the birth of agriculture and has reached its climax with the modern global timber industry. A central aspect of this chapter has been to try to reveal the contemporary economic processes that are leading to increasing levels of forest clearance in environmentally sensitive areas. Through an exploration of globalization and the multinational corporation we have seen how a desire to both reduce costs and increase the consumption of timber products has generated the economic pressures that are now leading to the large-scale illegal logging of protected forests in many less economically developed countries. These are the forces that Chut Wutty, who we heard about at the start of this chapter, was bravely fighting against in Cambodia. The key message of this chapter is that while the illegal deforestation of protected and ecologically significant woodland may seem like the acts of isolated criminals, they are actually the product of a global system of timber sourcing and manufacture that pursues low costs and exploits global economic competition.

At the end of this chapter, we briefly discussed the attempts that were being made to reform the global trade in timber through new corporate environmental strategies and wood certification procedures. These practices are being supple-

---

Find out more about the UN-REDD programme at:

http://www.un-redd.org/

---

mented by international policies designed to reduce emissions from deforestation and forest degradation (generally referred to as REDD initiatives). These initiatives attempt to create market values and payments that incentivize the protection of forest resources in order to help tackle climate change and to protect the indigenous groups who depend on woodlands. This chapter has illustrated that these initiatives are unlikely to be successful in the long run if they are not paralleled by broader reforms in the global timber industry. Ultimately, this will mean that people will have to accept rising prices for timber products and a consumer culture that is less devoted to the mass consumption of wood.

NOTE

1 Pasture woodlands are open-canopied forests – akin to landscapes that you can find today in the New Forest in England (Coelho, 2009).

KEY READINGS

Chapman, P. (2007) *Jungle Capitalists: A Story of Globalization, Greed and Revolution*, Canongate Books, Edinburgh. This volume provides a highly engaging account of the history of the United Fruit Corporation and its impacts on the rainforests of Central America.

Dauvergne, P. and Lister, J. (2011) *Timber*, Polity Press, Cambridge. This highly accessible volume provides an excellent introduction to all aspects of the global timber industry.

Ruddiman, W.F. (2005) *Plows, Plagues, and Petroleum*, Princeton University Press, Princeton, NJ. This book provides a detailed analysis of the impacts of agriculture on the global environmental system.

CHAPTER SIX

Cities

Sprawl and the urban planet

6.1 INTRODUCTION: URBANIZATION AND WHY DARWIN WAS WRONG AFTER ALL

In 2006 *The Guardian* newspaper's then Asia environment correspondent, Jonathan Watts (2006), made the bold assertion that the theories of Charles Darwin were soon to be proven wrong. The basis for his argument was the fact that by 2008 humanity would have collectively reached a geographical tipping point. This tipping point would occur when a new urban migrant, or the birth of a new metropolitan baby, would result in more people living in urban than rural areas. Indeed, by 2008 not only were some 3.2 billion people living in cities and urban areas, but the rate of expansion associated with the global urban population was accelerating. The United Nations estimates that approximately 50 million people (that is a similar number of people to those who currently live in South Korea) are added to the population of the planet's cities and suburbs every year (Flavin, 2007: xxiii). This growth is being increasingly concentrated in Asian cities. Following its 2011 census, for example, China announced that 51.3 per cent of its 1.3 billion population lived in urban areas. Figures also revealed that 21 million people moved to Chinese cities in 2011, and that there had been a 14 per cent increase in the number of people living in Chinese cities in the decade between 2001 and 2011 (BBC, 2012a). Jonathan Watts's point was that while Darwin suggested that the species that thrived most successfully were those who were able to adapt to their environment, the expansion of the human species had been based on the creation (in the form of cities) of environments that had been adapted to suit their own needs (for more on Darwinian thinking see Chapter 8).

The growing significance of urbanization throughout the world has important implications for human–environment relations. Urbanization is one of the Anthropocene's defining spatial characteristics. The significance of cities to the Anthropocene is perhaps expressed most clearly in two statistics. First, the United Nations Environment Programme estimates that 80 per cent of all anthropogenic carbon dioxide emissions are a product of 'urban-based activities' (UNEP/ Habitat, 2009). Second, the Organization for Economic Co-operation and Development claims that around 50 per cent of total public spending on environmental policies and services in more economically developed countries is spent by local, predominately metropolitan, governments (OECD, 2010). These two statistics reveal that urban areas are significant contributors to both environmental degradation and protection. When you also consider the important role that a series a key *global cities* (such as London, New York and Tokyo) plays in coordinating the spread and flow of global finances (Sassen, 1991), it is necessary to acknowledge the role that cities play as command and control centres for the Anthropocene. It is in this context that as we move through this chapter we consider the extent to which it may be helpful to consider our contemporary geological period not as the Anthropocene so much as the *Metropocene*: a period defined by the dynamics and demands of urbanization.

This chapter begins by providing a brief history of the urbanization process. In this section particular attention is given to the connections that exist between modern urbanization and the emergence of industrial society. The second section considers prominent frameworks that can be used to help us understand why urbanization is such a dominant force in the world today. The third section outlines the environmental consequences that are associated with urbanization. This section draws attention to the *external*, and increasingly global environmental impacts of urbanization, as well as to the *internal* impacts of rapid urban growth on the quality of more local metropolitan environments. The final section outlines the debates that surround the role of cities within the Anthropocene. These debates concern the extent to which cities, through their concentration of resources, population and technology, can provide solutions to the major environmental problems of the current era, or whether the rapid, unregulated expansion of cities is actually a major contributory factor to these problems.

6.2 A BRIEF HISTORY OF URBANIZATION: FROM MESOPOTAMIA TO THE MEGA-REGION

Like fridge freezers, microwave ovens and bullet trains, most major cities feel like very modern contemporary things. While cities are undoubtedly prominent hubs for various modernizing forces within society, they have actually been with us for much longer than you might think. If urbanization reflects the relatively large-scale concentration of

Read more of Jonathan Watts's work on environmental issues in Asia at:

http://www.guardian.co.uk/profile/jonathanwatts

people, resources and infrastructures into one place, the first incidence of urban areas dates back to the Neolithic age around 4000 BCE, when small cities started to emerge in the fertile plains between the Tigris and Euphrates rivers (see Figure 6.1). This area, which is now contained predominately in the modern state of Iraq, was known at the time as Mesopotamia (meaning 'land between two rivers'). While the early cities that emerged in this region, such as Ur, Uruk and Lagash, were very different from what we understand the modern city to be today, the reasons behind their emergence provide important clues to the forces that contributed to the rise of the urban form. Mesopotamia contained fertile soils, and with the use of irrigated water from the surrounding river systems, the region gradually saw the emergence of the first large-scale agricultural economy. It is what this large-scale agricultural society was able to achieve that proved to be so important to the birth of the first major cities.

Utilizing its natural geological and geomorphologic advantages, and the technological advances that were emerging in the Neolithic period (particularly metal working and irrigation), Mesopotamian society was able to produce a sustained surplus in agricultural production. This agricultural surplus was to have profound implications for the history and geography of human development. Simply put, an agricultural surplus means that an individual farmer, or a broader agricultural society, produces more food than is required for its immediate survival. This calorific surplus can be stored and saved in case less productive times affect the community in the future. When surplus production is, however, sustained over longer tracts of time, complex systems of trade and exchange can emerge. On the basis of these systems it becomes possible for certain segments of society to no longer have to concern themselves with the production of their own food.

In his classic 1961 analysis of the historical evolution of urban life, *The City in History*, the America historian Lewis Mumford highlighted the role that surplus agricultural production played in defining the form and function of cities. Mumford (1961: 29) observes:

In view of its satisfying rituals but limited capabilities, no mere increase in numbers would, in all probability, suffice to turn a village into a city. This change needed an outer challenge to pull the community sharply away from the central concerns of nutrition and reproduction: a purpose beyond mere survival.

Mumford essentially argues that cities are not, first and foremost, about a human desire to

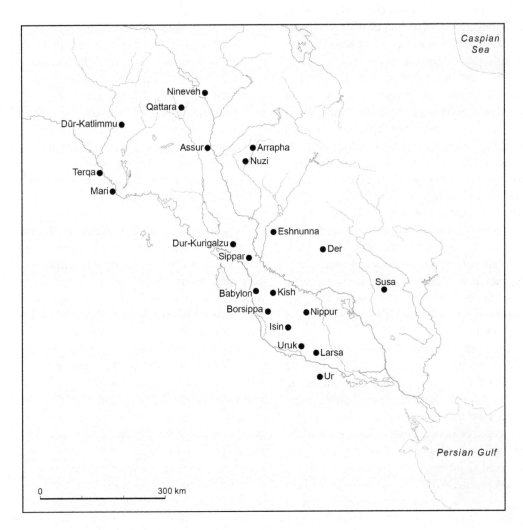

Figure 6.1 Settlements in early bronze age Mesopotamia
Source: Wikimedia Creative Commons

congregate in large numbers; instead they reflect a way of organizing society once the basics of human survival are routinely satisfied. The key to understanding the origins of cities is thus recognizing that they enabled people who had been freed from the all-encompassing responsibilities of food production, home building and reproduction to be engaged in more specialist activities. It is in this context that Mumford (1961: 32) states that with the rise of cities historians can see 'The expansion of human energies, the enlargements of the human ego, perhaps for the first time detached from its immediate communal envelope the differentiation of common human activities into specialized vocations'.

While urban centres have historically had different functions (from centres of trade, to sites of defence), Mumford argues that their defining function has been to enable the increasing specialization of human work and activity. By concentrating people and resources in one place, urban centres enable people to develop expertise (perhaps in carpentry or food processing, education or banking). People could specialize in such activities in cities for two reasons: 1) they did not have to concern themselves on a daily basis with food production; and 2) the dense and diverse population of urban areas contained varied skills, meaning it was possible for people to trade the skills they had for the goods and services others were able to offer.

The work of Mumford on the history of the urban form has two key points of general note for the themes running through this book. First, by connecting cities with human vocational specialization, Mumford makes a clear connection between the rise of cities and rise of the type of technologically driven, scientifically advanced civilization that is associated with the Anthropocene. Without the early city of Ur there could be no nuclear science; without Babylon no digital technology. Second, Mumford's work reveals the impacts that urbanization has had on human–environment relations. At one level, cities

disconnect people from the intimate relations that people had with nature in early agricultural and pre-agricultural societies. At another level, the types of specialist skills people developed in cities enabled them to acquire further command over the environmental resources that surrounded them. Ironically, perhaps, one of the first major impacts of urbanization on human–environment relations was the development of more advanced and ambitious agricultural techniques. Mumford (1961: 30) observes:

> This new urban mixture resulted in an enormous expansion of human capabilities in every direction. The city effected a mobilization of man-power, a command over long distance transportation, an intensification of communication over long distances in space and time, an outburst of invention along with a large scale development of civil engineering, and, not least, it promoted a tremendous further rise in agricultural productivity.

Despite the profound implications of early patterns of urbanization for human–environment relations, it was not until the nineteenth century that urbanization started to become a dominant force within human social, economic and environmental relations. With the industrial revolution taking hold in places such as Great Britain, France, Germany and the US, the nineteenth century witnessed a significant shift in population as many rural workers sought employment in the factories and workshops of the city (see Harvey, 1989a). With newly developed techniques for iron and steel production, significant advances in industrial chemistry and the impact of steam power on production techniques, cities were transformed from centres of specialist handicrafts to large-scale production systems (Mumford, 1961: 446). The impacts of this first round of industrialization on the nature and form of cities can be seen in the specific case

Box 6.1 Lewis Mumford

Lewis Mumford was an American historian and urban critic. He argued that the rise of the modern industrial city had been driven by greed and had seen urban planning prioritize the needs of economic development ahead of those of people and nature. Mumford was particularly critical of the processes of suburbanization. Within the suburbanization process he felt that the requirements of automobile travel had become the main focus of the urban planner. Mumford also argued that suburbs had the effect of alienating people from each other (as they now had to spend longer in the private space of the motorcar) and from the natural world (as suburbanization often meant that many in the city had farther to travel to reach nature). Through his writing on the city (which culminated in his epic 1961 tome *The City in History*) and his involvement with planning organizations (such as the Regional Planning Association of America) he called for the wiser, regionally based planning of urban development. He argued that by planning cities at a regional scale the needs of both urban development and nature could be more effectively balanced. Mumford also felt that modern cities could learn a lot from the urban forms that were common in medieval cities, where much more diverse forms of urban space existed and closer ties were apparent between the city and its surrounding environment.

Key reading

Luccarelli, M. (1995) *Lewis Mumford and the Ecological Region: The Politics of Planning*, The Guildford Press, London

---

See Lewis Mumford's short documentary film on the industrial city by searching for 'Lewis Mumford on the City' on YouTube.

of Manchester, England. In 1760, when England was still a predominantly rural-dwelling country, the population of Manchester stood somewhere between 30,000 and 45,000 people (Mumford, 1961: 455). By 1801 this population had risen to 72,000, and by 1851 to 303,382; that is nearly a sevenfold increase in the size of the city in the space of 91 years (Mumford, 1961: 455). Such patterns of rapid urbanization were replicated in other industrial centres such as Birmingham and would culminate in London becoming the first modern city to have an urban population in excess of 1 million people.

The next significant phase of urbanization would occur in the twentieth century and see the rise of a distinctively American style of urbanization. Unlike its densely packed European counterpart, the American city of the twentieth century was a city that was defined by the spatial expansion of the city. This new urbanization process was facilitated by the rise of mass transit systems (including both public transport and the motorcar), and led to the creation of a new urban phenomenon: sprawl (see Wolch et al, 2004). If London, Paris and New York were the quintessential industrial cities, it would be cities such as Los Angles and Las Vegas that would come to represent the new urban form (see Soja, 1989; Davis, 1999). Two main characteristics define sprawling cities such as Los Angeles and Las Vegas. First is their rapid pattern of urban growth. Taking Los Angeles as an example, as late as the 1870s the city was really no more than a small town, but between the 1880s and the 1920s its population rose from 35,000 to nearly 1 million inhabitants (Soja, 1989: 194). Following this 28-fold increase in the size of its population, Los Angeles continued to expand rapidly throughout the twentieth century, reaching a modern-day population of around 10 million people. The second key characteristic is the spatial size of these sprawling cities. The urban region that is now part of Los Angeles County, for example, encompasses 10,570km[2]; that means that the city region now occupies approximately the same amount of land as the whole state of Massachusetts.[1] To put things another way, between 1975 and 1990 Los Angeles's population grew by 45 per cent, but this increase in population was coupled with a threefold increase in the surface area of the city (Davis, 1999). In his book *The Geography of Nowhere*, James Howard Kunstler (1994) argues that the suburbanized city is an urban form that has only been made possible by the ready availability of affordable oil (for a broader discussion of the contemporary significance of oil to society see Chapter 2 in this volume). We discuss in greater detail the future challenges that appear to be facing such cities later in this chapter.

Los Angeles has grown to such an extent that it has now become part of a new breed of urban centres identified by the term 'megacity'. While various definitions of megacities exist, they are generally identified on the basis of having populations in excess of 10 million people. On the basis of this measure, there are currently 23 megacities in existence today (this figure is,

however, set to rise to 36 by 2025) (*Observer*, 2012: 36). While some of these megacities are industrial (the Rheine–Ruhr agglomeration) and suburban (Los Angeles) cities, which have expanded to megacity status over a relatively long time periods, many megacities of the present and future are still going through rapid episodes of urban growth. Much of this new wave of rapid urban expansion is occurring in South and Southeast Asia. If we take Indian urbanization as an example, it is estimated that by 2030 this single nation state will be home to six of its own megacities, with a combined total of 56 million people living in the cities of Delhi and Mumbai (*Observer*, 2012).[2] The increasing scale of urbanization that is associated with megacities, combined with their unprecedented rates of growth, are making it very difficult to effectively plan the provision of services and control the environmental harm caused within such places (see Marcotullio and McGranahan, 2007).

Although megacities may appear to reflect the ultimate form of urban growth, a new urban phenomenon appears to be now emerging. In southern China we are starting to see the first example of multiple megacities merging to form a mega urban region. In China, the cities of Hong Kong, Shenhzen and Guangzhou are beginning to amalgamate. This so-called 'endless city' has a combined population of approximately 120 million people. Other mega-regions are emerging in Japan between Nagoya, Osaka, Kyoto and Kobe (with an estimated combined population of 60 million people by 2015), and in Brazil between Rio de Janeiro and Sao Paolo (combined population of 43 million).

Los Angeles and the postmodern city

For more on the nature of urbanization in Los Angeles, and the nature of the so-called postmodern city, search for 'Ed Soja – The Postmodern City/Bonaventure Hotel' on YouTube. This short video is an excerpt from an Open University documentary that was shown on BBC television in the early 1990s.

What is clear is that urbanization is now the dominant way in which society organizes spaces of economic activity and social life. It is also apparent that the particular forms that urbanization take are becoming ever more varied and complex. To these ends, it is important that when we begin to analyse the environmental implications of urbanization that we do not see urbanization as a singular process. While often characterized by similar underlying processes (see next section), cities vary greatly in their spatial form and modes of operation (see Roy, 2009). Consequently, when you are considering older industrial cities such as Manchester and Birmingham, the sprawling low-rise cities of Los Angeles and Las Vegas, rapidly expanding megacities such as Lagos or Kinshasa, or mega-regions such as Hong Kong–Shenhzen–Guangzhou, you are likely to find very different forms of environmental relations and ecological problems associated with them.

6.3 THEORIZING THE CITY: FROM GROWTH MACHINES TO THE *FAVELA*

In the context of the rising significance of cities during the nineteenth and twentieth centuries, academic enquiries into the nature, form and function of cities gradually emerged. Chicago was home to one of the earliest and most prominent groups of urban scholars. The Chicago School of urban scholars (including Ernest Burgess, Louis Wirth and Robert E. Park) saw the city as a laboratory within which to study emerging social patterns and practices (see Bulmer, 1984). These studies led to the construction of the famous models of urban land use that continue to be widely taught in geography classrooms throughout the world. Despite its role as a point of origin for urban studies, this section is not concerned with the theoretical modelling of urban space that was characteristic of the Chicago School. This section instead focuses on a branch of urban theory that began in the late 1960s and early 1970s. These approaches were not concerned with mapping the internal dynamics of the city, but were instead interested in the connections that existed between capitalist economic and political systems and urbanization.

6.3.1 David Harvey and the capitalist city

In his celebrated book *Social Justice and the City* (1973), the British urban geographer David Harvey established one of the earliest frameworks for thinking about the connections between urbanization and capitalist economic development. Drawing on the insights of Marxist political economy, Harvey developed a *process-based* reading of the city. This process-based reading of the city rejected studies of cities that saw them as closed spatial systems, following an internally driven logic (a view developed within the Chicago School). Through a series of publications on the city throughout the 1970s, 1980s and 1990s, Harvey argued that cities constituted bundles of interconnected political and economic processes (including the production of goods and services, financial investments and the buying and selling of land) that connect urbanization to the logics of capitalism (see Harvey, 1985a; 1985b; 1989a; 1989b; 1996).

Following his appointment as an associate professor in the Department of Geography and Environmental Engineering at Johns Hopkins University, Baltimore, in 1969, Harvey was able to see close-up the dynamic ways in which capitalism and urbanization are connected. At the heart of Harvey's argument is the assertion that it is no coincidence that urbanization really started to emerge as a significant social and economic force at the same historical point that we see the rise of the first forms of capitalist industrial systems. For Harvey, cities are not merely patterns of settlement, they reflect the spatial logic of capitalist development. The relationship between cities and capitalism can, at one level, be seen in relation to what Harvey terms the *primary circuit of capital*

Plate 6.1 Urban structured coherence?
Source: Tim Cresswell

(Harvey, 1985a). The primary circuit of capital involves the production, circulation and trade of goods such as TVs, motorcars and food. Harvey noticed how cities effectively collected together the different ingredients that are needed for the production and circulation of goods. First, cities see factories and production centres located in close proximity to where large numbers of workers live. Second cities combine high concentrations of warehouses and distribution depots with the best road and rail infrastructure. Third, urban centres see the clustering of retail centres and shopping malls that sell products, alongside the popula-

tions and consumers who ultimately buy these products. This combination of factories, workers' housing, infrastructure and sales centres reflect what Harvey describes as a *structured coherence* within cities: a kind of space that has all of the necessary ingredients that are required for capitalism to thrive.

A key aspect of Harvey's analysis of urbanization is the argument that because cities are deeply interconnected with capitalist development, urban spaces are subject to the fluctuations and crises associated with capitalism. During the late 1960s and early 1970s, for example, a series of traditional blue-collar cities such as Baltimore, Birmingham and Detroit saw their structured coherence unravel as global capitalism entered a period of crisis and restructuring. In the wake of a global downturn in economic fortunes, many companies that had been located in these cities sought to preserve and enhance their profit margins by moving production to offshore locations where labour costs were greatly reduced. But what happened in these cities is indicative of another way in which capitalism and urbanization are interconnected. During the 1970s and 1980s, older industrial cities went through a period of major restructuring (what Harvey terms *creative destruction*) (see Harvey, 1989c). This restructuring saw new rounds of financial investment flowing into the property markets of these cities. This, so-called, *secondary circuit* of capital is associated with the flow of money into the acquisition of land and the construction of new urban offices, condominiums and housing. As the recent subprime housing crisis in North America has shown, this secondary circuit of urban capitalist investment is also crisis prone (for an excellent

Harvey talks about urbanization in Baltimore

In YouTube search for 'david harvey/city from below/part 2/urban crisis/amnesia/ reconstruction/promises'. This video shows David Harvey overlooking Baltimore's Inner Harbor and talking about changing patterns of urban development in the city.

Understanding urbanism through watching *The Wire*

A helpful starting point for thinking about the complex connections (of what David Harvey calls structured coherence) of a city is to watch the HBO series *The Wire*. Set in the city of Baltimore, a place that was central to the development of Harvey's Marxist interpretation of the urban form, *The Wire* reveals the ways in which various spheres of urban life connect together in the everyday life of this (admittedly dysfunctional) city. There is a strong focus in *The Wire* on the Baltimore Police Department and its attempts to investigate and control the narcotics industry within the city. According to its creator David Simon, however, *The Wire* is not a programme about the police, or the drugs industry per se, it is an account of how people live together in the city. To these ends, as the five series unfold you can follow the flow of money, power and influence from the drug corners of West Baltimore to the dock workers' unions, through lawyers' offices into the education system, from harbour-side developments to the city mayor and newspaper system. What emerges most clearly from this gripping and gritty account of urban life are ways in which the different sub-systems of the city are interconnected, and how events in one part of an urban network ripple throughout others. *The Wire* is also a powerful reminder of the fact that cities are shaped by the interplay of money and politics and that any attempts to reform a city cannot afford to ignore its existing political and economic structures.

Warning!

Before you begin to watch *The Wire* it is worth noting that it does contain strong violence and language throughout.

---

overview of Harvey's work on cities, see Merrifield, 2002: 133–155).

6.3.2 The city as a growth machine

While the second circuit of capitalism may have offered a form of second life to many older industrial cities (albeit a relatively short one), for some cities property-based development has always been their primary economic dynamic. In a paper published in 1976 and entitled *The City as a Growth Machine*, Harvey Molotch described how cities such as Los Angeles, Las Vegas and other *sunbelt cities* have developed a pattern of urban economic development that is based on property development (Molotch, 1976). Many sunbelt cities are characterized by high-skilled, white-collar employment in the financial, high-tech and bio-

medical sectors (see Krueger and Gibbs, 2008). The fact that such workers tend to prioritize larger forms of suburban housing, with associated land, means that there is a natural impetus behind the spatial expansion of such cities. But according to Molotch there is a deeper, capitalist logic to urban sprawl. Molotch argues that the real estate sector benefits significantly as more and more land is annexed by a city and made available for development. Essentially, as a city grows, what may have previously been worthless land in the middle of the desert becomes highly valuable real estate. The spatial growth of the city essentially increases the amount of land that has potential commercial value and can be subject to speculative financial investment. Molotch's theory of the *urban growth machine* recognizes the benefit that urban spatial expansion brings to various sectors in the city.

Plate 6.2 Suburbanization in Toronto
Source: Author's own collection

For example, while you might expect to see urban planning authorities objecting to the sprawl of cities (perhaps on the basis of the extra pressure that such developments place on the provision of roads, sanitation and other basic utilities), metropolitan authorities also benefit from the spatial expansion of cities in the form of increases in their tax base and associated sources of public revenue. Local newspapers could offer resistance to the growth of cities on the basis of its impacts on traffic levels in the city and the rise of associated forms of air pollution. But, of course, urban newspapers also benefit from the growth of cities as their potential readership and circulation area increases. Molotch thus argues that rapidly sprawling cities tend to be characterized by a *coalition of interest* that converges on the theme of *growth*. Given the collective power of such coalitions of local landowners, politicians, planners

and newspaper editors, it becomes very difficult to stop the unregulated expansion of cities, despite the adverse environmental effects this may have (see next section). It is in this context that Smith (2002) claims that for many cities now the second circuit of capitalism is actually the principal circuit of capital accumulation (see also Roy, 2009).

6.3.3 Alternative interpretations of the city

The dynamics of urbanization identified by both Harvey and Molotch are helpful when it comes to trying to understand the environmental impacts of urbanization and what may be driving them. However care must be taken not to reduce the complexities of the urbanization process to either the dynamics of industrial production or property-based growth (see Castells, 1983; Ward, 1997).

It is also important to keep in mind that much contemporary urban theory has been based upon research carried out on cities in North America, Western Europe and Australasia. Given that the most rapid forms of urbanization (particularly those associated with megacities and mega-regions) are now occurring in South and South-east Asia and Africa, it is important to develop theories that can allow for the particularities of urban development in these places. Ananya Roy has drawn particular attention to the importance of developing new understandings of urban development that are able to deal with the complexities of urbanization in less economically developed countries (Roy, 2009). Drawing on observations of urbanization in the global south, Roy identifies a series of distinctive processes that characterize city development there. These characteristics include: the impacts of the colonial past, the unprecedented pace of urbanization, the significant role of urban development in state formation and consolidation, the role of informal economic practices, and, perhaps most significantly, the *informal production of space* associated with the rise of shanty towns (Roy 2009: 826). It is important not to underestimate the significance of informal patterns of urban development. According to Davis (2007), slums and *favelas* (i.e. shanty towns) are not temporary urban accidents. With some 200,000 slums now in existence worldwide, Davis argues that they are becoming an increasingly important blueprint for urban development. What the work of Roy ultimately suggests is that the increasingly improvised form which urban development is taking in the global south (ranging from the building of shelter to the provision of services) requires new ways of interpreting the city (for more on issues of improvised adaptation see Chapter 9).

6.3.4 The environmental blind spot in urban studies

An important aspect of urban theory that has, so far, not been mentioned pertains to environmental concerns. For a significant period of time there was something of an *environmental blind spot* within urban theory (see Harvey, 1996; Light, 2001). This blind spot was in part a product of a division that emerged between socio-economic studies of cities as predominately human entities, and environmental studies of those places outside of cities that constituted nature. The unhelpful divide between urban research and environmental studies has been gradually bridged by a series of studies that have explored the ways in which cities are deeply implicated in environmental systems at a range of different scales. It was actually students of David Harvey such as Neil Smith and Erik Swyngedouw who conducted much of the early work in this field (see Smith, 1984; Swyngedouw, 2007). This was supported by David Harvey's own analysis of the urbanization of nature in his 1996 volume *Justice, Nature and the Geography of Difference*. Working from a Marxist-based reading of the city, and studying issues related to urban water supply, forestry, air pollution and food production, this research has exposed the complex ways in which cities transform and metabolize nature (for an overview of this area of enquiry see Heynen et al, 2006). This rapidly expanding area of work is now referred to as *urban political ecology*. Just as David Harvey attempted to develop a process-based reading of the city, which connected urban development to the flows of capital, urban political ecologists extend this approach in order to consider how cities channel the flows of capital investment, housing construction and the production of goods alongside the flow of environmental resources such as air, water, foodstuffs, minerals and energy (see Keil, 2005).

6.4 URBANIZATION AND THE ENVIRONMENT

Having explored the history and nature of urbanization, in this section we move on to explore the connections that exist between cities and the environment. This task is split into two sections. In the first section we explore the different ways in

Plate 6.3 The informal production of urban space – Rocinha *favela*, Rio de Janeiro, Brazil
Source: Wikimedia Creative Commons

which cities are involved in the transformation of the environment at a range of different scales. In the second section we consider the debates that surround the nature of urban–environment relations, and, in particular, the extent to which cities are positive or harmful influences on the ecological systems upon which humans depend.

6.4.1 Cities and environmental transformations

In his account of the environmental history of Chicago, *Nature's Metropolis*, Cronon (1991) provides a detailed analysis of the ways in which urban development (at least within older industrial cities) is based upon the expanded exploitation and transformation of nature. Chicago is located in the American Midwest, a midway point between the old colonies of the eastern seaboard and the western frontier (see Figure 6.2). A central part

of Cronon's argument is that while conventional accounts of the western frontier tend to portray American history as the story of the taming of nature along a gradually progressing line of civilization, this history is actually better described as a process of urbanization. For Cronon, the city that lay at the centre of this process was Chicago. Chicago's location meant that it was ideally placed to exploit the environmental resources that were to be found in the heartland of the North American continent. The American Midwest is home to highly productive woodlands, fertile prairies and lush rangeland pasture. Furthermore, Chicago is located on Lake Michigan (part of the interconnected Great Lakes water system), which meant that it was in an advantageous position when it came to the logistics of transporting and trading goods.

For Cronon, the history of urbanization in Chicago is inextricably linked to the ways that

Figure 6.2 Chicago and the Great Lakes

the city was able to exploit its environmental hinterland in the Midwest. At one level the connection between Chicago and its surrounding environment can be seen in the case of lumber (Cronon, 1991: 148–206). The early expansion of Chicago was reliant on the exploitation of the timber resources that were to be found in the white pine forests to the north of the city (see Chapter 5 of this volume). Millions of tons of felled trees were transported into Chicago during the nineteenth century. This flow of timber was used in the construction industry within the city. The exploita-

tion of woodlands also enabled the establishment of a lumber district in Chicago, which specialized in the production and distribution of timber products (Cronon, 1991).

Urbanization in Chicago was further connected to the environments of the American Midwest through the emerging agricultural industry. During the nineteenth century, Chicago developed one of the earliest industrial-scale meat processing industries (Cronon, 1991: 207–259). As the so-called 'Great Bovine City of the World', Cronon describes how Chicago was able to '[a]ssemble the animal products of the Great West, transmute them into their most marketable form, and speed them on their way to dinner tables around the world' (Cronon, 1991: 211). In addition to pigs and cattle, Chicago also became a crucial processing, storage and market place for grain (Cronon, 1991: 97–147). As a specialized centre for the collation and transformation of agricultural produce, the growth of Chicago into a major metropolis was deeply interconnected with the wholesale transformation of the soils and ecologies of the American Midwest.

Cronon's analysis of Chicago provides us with a very detailed analysis of the (often hidden) connections that exist between urbanization and environmental change. Moving beyond the specific example of Chicago, however, it is helpful to think of the relationships that exist between cities and the environments on which they depend in more general terms. The environmental Kuznets' Curve offers a helpful starting point when trying to understand urban–environment relations. The environmental Kuznets' Curve is the bell-shaped

curve labelled 'metropolitan scale' in Figure 6.3 below. What this curve demonstrates is a generalized pattern of urban environmental relations over time (see Marcotullio and McGranahan, 2007; Marcotullio, 2007). The general trend suggested by this curve is that early forms of urbanization (particularly in industrial cities) tend to be characterized by increasing rates of environmental degradation at a local scale (measured in air pollution, resource use, water quality). At some point, the curve suggests that cities pass through some kind of transitional zone, after which, and following a sustained period of economic growth, the levels of local environmental degradation associated with urbanization tend to decline. While the environmental Kuznets' Curve is only a generalized model of urban environmental relations, the pattern it charts reflects the experience of many industrial cities.

There remains some debate concerning precisely what causes cities to go through a period of transition in their environmental relations. It could be that environmental improvements come

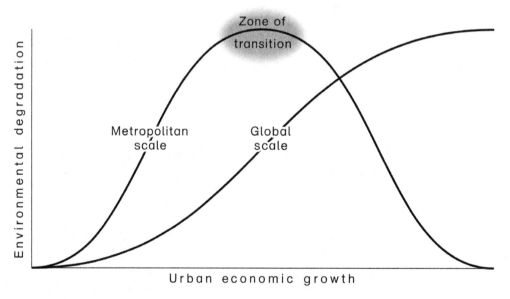

Figure 6.3 The (urban) environmental Kuznets' Curve and the impacts of cities on global environmental issues

Source: Adapted from Marcotullio and McGranahan, 2007

as a result of the growing power of municipal governments who are able to more tightly regulate environmental degradation through laws and planning restrictions (see Whitehead, 2009). Another explanation could be that as urban economic wealth increases a new, affluent urban middle class emerges that starts to demand improvements in the environmental relations of the city. A third explanation suggests that at certain levels of economic growth, cities are able to invest a significant amount of resources into technological solutions to the causes of environmental pollution (including more efficient energy generation systems, modern mass transit systems, better fuel efficiencies in motor vehicles and the building of waste treatment plants) (Weale, 2001). A final explanation suggests that as they develop, cities become adept at outsourcing polluting industrial activities, which characterized their early phases of development, to lower cost locations in other parts of the world (for an overview of these theories see Marcotullio, 2007).

This final point leads us into a discussion of the second curve depicted in Figure 6.3. This curve (labelled 'Global Scale') depicts the increasing contribution that cities appear to make to global forms of environmental degradation as their wealth increases. These globally scaled forms of environmental degradation can take many different forms. The most obvious form these take are the rising levels of carbon dioxide that are emitted from cities and contribute to changes in the nature of the global climate system (see Chapter 3). As cities grow and develop what Kunstler (1994) has called the auto-suburb, their greenhouse gas emissions tend to rise as a result of increases in automobile usage and average car journey lengths. But the increasing carbon footprint of cities is not only a product of the greenhouses gases that are emitted from their commuter suburbs. As post-industrial cities see an increasing amount of the products that they depend upon manufactured elsewhere, we tend to see rising levels *of embedded carbon dioxide emissions* associated with such urban centres.

Embedded carbon dioxide emissions reflect the carbon dioxide releases associated with the products that are produced in one place but consumed in another. To these ends, a significant portion of the carbon dioxide emissions associated with a post-industrial city's consumption needs is actually emitted in distant, low-cost production centres. These, often overlooked, emissions do, however, reflect an important aspect of the global environmental relations of many cities (see Chapter 1 for a broader discussion of notions of relational space).

The carbon footprint of cities reflects one way in which urban centres contribute to transforming the operation of global environmental systems. Other forms of globally scaled environmental degradation associated with urbanization tend to be less about the changing of global environmental systems and more about the global spread of environmentally damaging substances. A controversial example of globalized environmental pollution is evident in the practices of toxic colonialism (see Harvey, 1996). Toxic colonialism is a term that is used to refer to the transport and disposal of hazardous waste products in less economically developed countries. The toxic substances involved in this process include, among other things, nuclear waste, dioxins, poisonous metals, persistent organic pollutants and sanitary waste, which can all have harmful effects on both ecological systems and human health (FAO, 1999; Bernstorff and Stairs, 2000). The trade in toxic substances has recently taken a new turn with the rise of e-waste being transported to less economically developed countries. E-waste can take many different forms including unwanted mobile phones, video games consoles, TV sets and computers (see Koné, 2010). Recent estimates indicate that approximately 500 shipping containers of unwanted e-waste (the equivalent of around 100,000 computers and 44,000 televisions) is transported to Nigeria every month (*Ghana Business News*, 2009). The socio-ecological problems associated with this e-waste trade emerge when the products are disassembled for recycling.

Plate 6.4 E-waste collection in the UK
Source: Author's own collection

Disassembly is associated with the release of lead, mercury and chlorinated dioxins, which are again dangerous to human health and the environment (*Ghana Business News*, 2009). The reasons for this international toxic exchange are predominantly economic: the treatment of toxic waste in Africa costs approximately $40 per tonne, compared to $2000 per tonne in more economically developed countries (Koné, 2010). When it comes to the global environmental relations of cities, this trade in e-waste is particularly telling. It is, of course, products such as computers and mobile phones that provide the high-tech infrastructures upon which the service industries of post-industrial cities depend. Their low-cost disposal in less economically developed countries is emblematic of how effective modern cities have become at redistributing their unwanted products downstream and downwind.

The environmental Kuznets' Curve provides us with a helpful context in and through which to consider the changing nature of urban–environment relations. Some have been critical of the environmental Kuznets' Curve because of the ways in which it suggests a positive relationship exists between urban economic development and environmental protection (at least in the long term). Others have questioned the extent to which the curve can be applied to the rapidly paced forms of urban development that are now being experienced in less economically developed countries (where rapid increases in environmental degradation may outpace the ability of governments and technology to combat them) (see Marcotullio, 2007). When the environmental Kuznets' Curve is combined with an appreciation of the increasingly globalized nature of urban–environment relations it becomes possible to see how just as cities become more prosperous, clean and verdant, their global ecological footprints may actually be increasing. While the work of Cronon (1991) is helpful in revealing the ways in which Chicago's early development was based upon the exploitation of its surrounding environmental hinterland, it appears that within the Anthropocene the hinterlands upon which cities depend are becoming increasingly global in nature.

6.4.2 Interpretations of cities in the Anthropocene

The complex nature of urban–environment relations depicted in the Kuznets' Curve is actually symptomatic of debates concerning the impacts of urbanization on human environmental affairs. At one end of the spectrum are a group we could describe as the urban optimists. This group is composed of a series of prominent urban thinkers and commentators such as Peter Hall (2003) and Herbert Girardet (2006) who argue that cities offer a basis for developing a more sustainable environmental future. The urban environmental optimists argue that the high concentrations of people and infrastructure that characterize cities mean that they provide an ideal spatial template for developing a low-energy context for human activity. It is consequently claimed that if urbanization is properly planned, cities offer unique opportunities to spatially coordinate the places where people live, work, go to school, recreate and shop (see Hall, 2003). These processes of urban planning are often described as *smart* or *new urbanism* and involve combining mix-used planning with investment in clean modes of public transport in order to produce a form of low-energy urbanism (Krueger and Gibbs, 2008).

At the other end of the spectrum are the urban environmental pessimists. These urban environmental pessimists, such as the environmental economist Joan Martinez-Alier (2003), are critical of the capacity of cities to provide well-ordered environmentally benign patterns of development. According to Martinez-Alier, so long as cities

remain dependent on the competitive exploitation of non-renewable resources, they will be unable to contribute to enhanced forms of environmental sustainability. To put things another way, while cities remain locked into the competitive drive to capture non-renewable energy – in order to produce more goods and services than their competitors and to open up more land for suburban commercialization – they will always be ecologically short sighted and stupid!

In reality of course, the situation is more complicated than either the urban pessimists or optimists appear to suggest. Different types of city, be they old industrial, low-rise suburban, megacities or global financial centres, have very different types of environmental relations (see While et al, 2004). As Table 6.1 illustrates, the political and economic nature of different cities means that they are able to support some environmental priorities, but tend to resist others. Consequently, while global financial centres may find it relatively easy to support carbon-reduction strategies in the high-tech workplace, they may be less inclined to oppose the airport expansions their forms of urban development appear to require. Likewise, property-based urban development strategies may find it beneficial to promote urban greening and the improvement of urban air quality, but they find it more difficult to promote less car use and reduce associated levels of carbon dioxide emissions.

A final, and perhaps, most worrying interpretation of the connections that exist between cities and the environment in the Anthropocene is offered in the work of Hodson and Marvin (2009). Hodson and Marvin argue that in the context of shortages in the availability of non-renewable resources, and the threats of climate change, we may be entering a new period in the history of urbanization. This new form of urbanization goes by the name of *urban ecological security*. According to Hodson and Marvin, this new period of urban ecological security is seeing powerful cities such as London, New York and Paris utilizing emerging environmental threats as

Table 6.1 The environmental priorities supported and resisted within different forms of city

Urban economy trajectory	Example cities	Environmental priorities supported	Environmental priorities contested
Rapidly expanding, oil-based urban economies	Lagos Khartoum Caracas	Urban greening	Polluter-pays initiatives Carbon reduction imperatives Energy descent strategies
Declining, heavy industrial urban economy	Katowice Sofia	Environmental clean up of contaminated land Enforcement of tougher air and water quality standards	Improving workplace environments Pollution taxation
High-tech, knowledge-based urban economies	Boston Cambridge San Francisco	Investment in environmental services connected to improved quality of life (green open spaces, improved air quality)	Reduced dependence on air transport and airport expansion
Centres for consumption-based services and entertainment	Las Vegas Dubai	Innovative eco-architecture Quality of life infrastructure Water conservation	Compact-living and anti-sprawl measures Reduce, reuse and recycle philosophies Energy efficiency
Global financial centres	London Frankfurt Sydney Tokyo New York	Climate change initiatives and carbon rationing	Restrictions on environmentally harmful international investments Airport expansions
Boomburbs and property-based urban economies	Austin Mesa	Quality of life infrastructure	Compact-living and anti-sprawl measures Congestion charges and anti-road traffic measures

Source: Mark Whitehead

the basis for forging a new competitive landscape upon which urban development can be based. By sharing new technologies and exploiting collective investment opportunities, such cities are attempting to be ahead of the game when it comes to being able to 'to *anticipate systematically and prepare strategically* for a period of [environmental] constraint' (Hodson and Marvin, 2009: 199). Such processes of urban anticipation and preparedness are associated with cities attempting to seal themselves off from the worst effects of climate change, while also securing a long-term supply of the energy that their city and its residents are likely to require. To these ends, urban ecological security is about the search for competitive advantage in the Anthropocene.

6.5 CONCLUSIONS

In this chapter we have achieved four primary things. First, we have established that urbanization is a prominent force within the contemporary era, and that cities are becoming an increasingly dominant context for living on the planet. Second, we have considered the historical evolution of cities and illustrated how their emergence was connected to the agricultural transformation of nature, and how the specialist practices and trades that have emerged within cities have played a crucial role in the subsequent transformations of the environment. Third, we have explored a series of theories that help us to understand the forces that are driving contemporary forms of

Box 6.2 Smart and new urbanism

Smart and *new urbanism* are two interconnected, but distinct, urban planning responses to the problems of suburbanization. Both planning movements share a desire to construct more interesting, walkable and community-oriented forms of urban development. To these ends, both approaches oppose the construction of bland and homogenous suburban landscapes, which are car oriented and provide little opportunity for social interaction. They do differ, however, in important ways. While new urbanism tends to prioritize the use of creative and diverse architectural styles in urban design (often with an emphasis on the neoclassical), smart urbanism tends to focus more on the goal of high-density, mixed-use planning. New urbanists tend to be critical of smart urbanism due to its failure to recognize the importance of architectural form in the creation of convivial metropolitan spaces, while smart urbanists suggest that new urbanism is more concerned with design aesthetics than the practicalities of planning liveable urban neighbourhoods. In reality many new and smart urbanist developments combine elements of both planning schools.

Plate 6.5 Dicken's Heath – A *new urbanist* development near Birmingham, UK
Source: Author's own collection

Key reading

Krueger, R. and Gibbs, D. (2008) 'Third wave sustainability: Smart growth and regional development in USA', *Regional Studies* 49(9): 1263–1274

FROM THE BLOG

Placing the Anthropocene: league tables and urban environmental competition

Posted on 8 October 2011 by mswaber

You may be interested in this recently produced league table of air pollution that has been reported in Forbes: http://www.forbes.com/sites/williampentland/2011/10/01/worlds-most-dangerous-places-to-breath/. In many ways league tables like this fit perfectly with the logic of what Bernstein (2000) has described as *liberal environmentalism*: create a competitive landscape for environmental innovation and the places that produce the best quality environment will benefit as people vote with their feet and move to them. These arguments have recently been articulated in Matthew E. Kahn's (2010) liberal environmental manifesto *Climatopolis: How Cities will Thrive in a Hotter Future*. According to Kahn (a direct descendent of the Chicago School of Economics) climate change represents not so much a threat but a new competitive landscape upon which 'green entrepreneurs' can carve out a new ecological circuit of capital accumulation (presumably to ease the crisis-ridden primary and second circuits associated with consumable goods and property, respectively). This cornucopian vision of the environmental future necessitates a global market place (to ensure that investment can flow en masse to the most environmentally innovative places) and continued economic growth (to ensure that the incentive for innovation is secured by the promise of wealth).

There are clear problems with the liberal environmentalism embodied in the work of Kahn and expressed in environmental league tables. At one level, they say little of who is able to move between places (largely a group of what Cresswell (2006) calls the 'kinetic elite'), or of how wealthy communities tend to displace polluting activities downstream, downwind and increasingly offshore. They are also based upon the assumption that technologies can/will solve our current environmental problems. It is not that I don't believe in the power of technological innovation, but rarely are technological developments produced in direct response to socio-ecological needs, or wielded by those who most need them. In a recent article for the *International Journal of Urban and Regional Research*, Hodson and Marvin (2009) describe the emergence of a new era of urban ecological security. Urban ecological security describes the processes in and through which already powerful places (such as New York, London and Tokyo) are using their wealth to create enclaves of security within which they can ward off the worst effects of climatic threats and energy insecurities. It seems to me that liberal environmentalism is much more likely to lead to the production of an archipelago of environmental privilege than a promised land of ecological security for all. Liberal environmentalism is built upon the necessary abandonment of certain (often disadvantaged) places as part of the uneven development of a new ecological circuit of capital accumulation.

Key readings

Hodson, M. and Marvin, S. (2009) 'Urban ecological security: A new paradigm?', *International Journal of Urban and Regional Research*, 33(1): 193–215
For more Blog posts go to: http://anthropocenedotcom.wordpress.com/

urbanization. Drawing on the work of David Harvey and Harvey Molotch we have established the ways in which urbanization reflects a form of spatial logic to capitalist forms of development. Fourth, and finally, we have considered the forms of environmental relations that characterize cities. In this context, it was argued that different forms of cities are presented with different forms of opportunities and constraints when it comes to reforming their environmental relations.

The most important insight of this chapter is that it is very difficult to understand what is happening in the Anthropocene without an effective account of urbanization. Furthermore, by revealing the ways in which cities are locked into broader processes of capitalist economic development, this chapter also illustrates that attempts to reform urban development (perhaps by stopping the suburban sprawl of cities or improving their public transportation systems), and reduce cities' ecological footprints, are not simply a technical planning issue. Changing the environmental relations of cities is a process that necessitates rethinking the logics of capitalism and the competitive imperatives of growth-oriented economic development. The contemporary attempts to develop regimes of urban ecological security, discussed in the final section of this chapter, suggest that contemporary environmental threats (such as climate change and energy shortages) may actually be the basis for a new round of competitively orientated urban growth and development, and not for a more ecologically benign period of urbanization. This is a process that is unlikely to solve the ecological challenges that lie ahead, and will most likely result in many of the poorest cities in the world being most exposed to the risks that lie in our collective environmental future.

NOTES

1 http://www.simonkelk.co.uk/sizeofwales.html. This comparative calculation was based on figures produced at the website SizeofWales. com.
2 These figures are based upon estimates of McKinsey and Co Consultants and were reported in the *Observer*.

KEY READINGS

Cronon, W. (1991) *Nature's Metropolis: Chicago and the Great West*, W.W. Norton and Company, London. This detailed and extensive environmental history of Chicago provides a fascinating insight into the relations that emerge between cities and the environment as urban centres grow and develop.

Harvey, D. (1996) *Justice, Nature and the Geography of Difference*, Blackwell, Oxford. An advanced text that provides an insight into Marxist urban theory and how such theories can be used to analyse urban–environment relations.

Kunstler, J.H. (1994) *The Geography of Nowhere: The Rise and Decline of America's Man-Made Landscape,* Free Press, New York. An engaging and accessible discussion of the rise of the suburbs and their socio-ecological consequences.

Marcotullio, P.J. (2007) 'Variations of urban environmental transitions: The experiences of rapidly developing Asia-Pacific cities' in Marcotullio, P.J. and McGranahan, G. (eds) *Scaling Urban Environment Challenges: From Local to Global and Back*, Earthscan, London: 45–68. This chapter provides a detailed discussion of the urban environmental Kuznets' Curve.

PART TWO

Living in the Anthropocene

CHAPTER SEVEN

Governing the environment

7.1 INTRODUCTION: PROTECTING PEOPLE FROM NATURE OR PROTECTING NATURE FROM PEOPLE?

According to the US National Hurricane Centre, Tropical Depression 12 formed over the southeast Bahamas on 23 August 2005 (Knabb et al, 2005). Six days later, Tropical Depression 12 made its final landfall near to the mouth of the Peale River close to the Louisiana/Mississippi state border. By this point it was no longer a tropical depression, but a category 3 hurricane going by the name of Katrina (Knabb et al, 2005). According to official estimates, Hurricane Katrina was one of the five most deadly hurricanes to strike the US, and the most costly in terms of the overall damage that it caused (Knabb et al, 2005: 1). The total cost of the damage has been estimated at $108 billion and 1833 deaths have been directly attributed to the effects of Katrina (largely concentrated in Louisiana, where there were 1577 fatalities) (Knabb et al, 2005: 11–13). While the events surrounding Hurricane Katrina have a number of significant implications for those studying socio-environmental relations, I would like here to briefly focus on one aspect of the disaster: what it tells us about the relationship between nation states and environmental government.

In the immediate aftermath of Katrina *The Economist* stated:

Since Hurricane Katrina, the world's view of America has changed. The disaster has exposed some shocking truths about the place: bitterness of its sharp racial divide, the abandonment of the dispossessed, the weakness of critical infrastructure. But the most astonishing and most shaming revelation has been of its government's failure to bring succor to its people at their time of greatest need. (*Economist*, 2005a: 11).

The crucial point to note here is that the direct impacts of the hurricane were far less severe than had been previously feared. Katrina had been downgraded from a category 5 to a category 3 hurricane by the time it left the Gulf of Mexico and entered Louisiana. Furthermore, the centre of Katrina's strongest winds and heaviest rains missed the heavily populated areas in and around the city of New Orleans (passing to the east of the conurbation). The problems that Katrina generated were to these ends, as much a product of human (in)action as they 'were acts of nature'. The floods that followed Katrina were a result of the failure of the levee system that had been designed to protect the urban populations in New Orleans from the water held in Lake Pontchartrain and the Mississippi River. It was the Army Corps of Engineers who had originally built the 350 miles of levees that surrounded New Orleans

(Handwerk, 2005). The levees failed in part because of poor general maintenance, but also as a product of the simple fact that they had never been designed to withstand a storm of Katrina's strength. If the levees' failure was the responsibility of the federal state, the slow response to the floods once they occurred was a failure of government at all levels. The hurricane evacuation plan that had been put in place for events like Katrina did not provide sufficient transport for those who did not own their own motorcar. Furthermore, FEMA (the Federal Emergency Management Agency) was slow to deliver the food, water and medical assistance needed by those who were left in New Orleans and who had to congregate in over-crowded, temporary accommodation such as the Superdome. Unfortunately, it was the poorest within the city of New Orleans, and particularly the African American community, who suffered most as a consequence of such failures (see Bullard and Wright, 2009). Further criticism has been levelled at the federal government's longer term response to Katrina. Many have suggested that the redevelopment strategies have been used as a basis for privatizing a range of New Orleans's public utilities and for awarding lucrative state contracts

Plate 7.1 The aftermath of Hurricane Katrina
Source: Wikimedia Creative Commons

to companies who had close connections to the George Bush Junior Administration (see Klein, 2007: 3–7).

In many respects, the most serious problems associated with Hurricane Katrina were not caused by the physical impact of the hurricane itself, but related technological and political failures within national, state and city governments. This chapter is interested in the role of government, particularly at a national scale, in the management and regulation of environmental affairs. Nation states remain a key context within which political and economic life is organized in the world today. The gradual emergence of nation states corresponds closely to the rise of the kinds of socio-environmental relations that have become associated with the Anthropocene. A key goal of this chapter is thus to uncover the dual role of states as both guardians of the natural world and as facilitators of the exploitation of the environment for human ends. While there is no direct evidence connecting Hurricane Katrina with the broader processes of global warming, it is clear that, in theory at least, these forms of extreme weather are likely to become more frequent and more intense if climate change continues on its current trajectories (see IPCC, 2007: Para. 3.8.3). It is in this context that we can immediately see the complex challenges that confront the state in its attempts to regulate human relations with the environment. While the US government may have resisted significant action on climate change for many years (due to the potential impacts of such action on its economy), one consequence of failing to protect the global environment could be the need to better protect its citizens from more frequent extreme weather events. This situation was summarized well in a piece that *The Economist* ran several months after the events of Katrina. It observed that 'after Hurricane Katrina, the balance between protecting people from nature, and protecting nature from people, has become an urgent matter of public policy' (*Economist*, 2005b: 29).

This chapter begins by exploring the historical emergence of the nation state and systems of large-scale government, and the impacts that this process had on the human exploitation and management of the environment. The second section outlines a series of theories that attempts to explain the role of states within the control and regulation of environmental affairs. The following two sections consider two case studies of state–environment relations: the first being the British government's reaction to the 1952 London fog disaster, the second being the interventions of the US federal government in the Florida Everglades.

When the Levees Broke

To find out more about the political controversies surrounding Hurricane Katrina, watch Spike Lee's four-part, HBO documentary about the hurricane, *When the Levees Broke: A Requiem in Four Acts*. A full length version of this documentary is available on YouTube at:

http://www.youtube.com/watch?v=lqCQ VVvNASE

7.2 A BRIEF ENVIRONMENTAL HISTORY OF THE NATION STATE

7.2.1 Unpacking and defining the state

The story of the nation state takes us back to 1648 and the towns of Osnabrück and Münster, in what is modern-day Germany. Between the May and October of 1648 a number of peace treaties were signed in these towns. These treaties sought to find a peaceful solution to a series of conflicts that had been raging for decades in Europe. These treaties, which collectively produced something that is routinely referred to as the Peace of Westphalia, are of historical significance because they laid the diplomatic and legal foundations of what we today refer to as the nation state. Before the Treaties of

Westphalia the political map of Europe was a complex amalgam of city-states, commonwealths, empires and kingdoms, which often had ill-defined and highly contested boundaries. The key outcome of the Peace of Westphalia was to define the principle of the sovereign state. A sovereign state is a political community with a clearly demarcated territory, which has the ability to determine its own internal affairs. While the nature and form of the sovereign state system has been through many forms since the Treaties of Westphalia, it has become the dominant way of organizing political life throughout the world.

The evolution of the nation state system in Europe

An interesting way of observing the evolving and ever-changing nation state system in Europe is to watch the graphic *Political Borders of Europe from 1519 to 2006,* which can be found on YouTube.

Before moving on to consider the relationship between states and the environment, it is important to provide some more detailed definitions of precisely what the modern nation state is and does. To these ends, the sociologist and historian Michael Mann (1984: 185) defines the nation state as:

> [a] differentiated set of institutions and personnel embodying centrality in the sense that political relations radiate outwards from a centre to cover a territorially demarcated area over which it exercises a monopoly of authoritative rule-making, backed up by a monopoly of the means of physical violence.

In Mann's definition, the idea of the nation state as a *territorial* entity is accompanied by an appreciation of the ways in which states tend to *centralize* power within a series of prominent locations (such as capital cities), be made up of series of specialized *institutions* (including parliaments, courts and government departments), and have control over the legitimate use of force (in the forms of the military and police authorities). While definitions like the one provided by Mann are helpful in enabling us to identify nation states and differentiate them from other forms of political institution, they can be misleading. While nation states are territorial entities, few would claim that their power stops at the borders' edge. In a related sense, while nation states are clearly defined by a set of identifiable central institutions, it is very simplistic to believe that power simply radiates out from these points to uniformly affect what occurs within a given national territory (Allen, 2003). Finally, while such definitions may tell us what nation states are, they actually reveal very little about the underlying purposes of state systems (for a more detailed overview of these arguments see Whitehead, 2008).

The work of the French philosopher Michel Foucault casts light on the deeper purposes of the modern state. According to Foucault, the contemporary role of the nation state is closely tied to the practices of *government.* For Foucault, governing involves a sense of care within the operation of states; a care that is directed at national populations, and seeks to ensure that society is ordered in such a way that it can function effectively in relation to the economic production of wealth (Foucault, 2007 [2004]). It is important to note that the practices of governing identified by Foucault predate the formation of nation states in the seventeenth and eighteenth centuries (Foucault actually traces such systems of collective care back to the early nomadic Hebrew nation). It is also clear that governing has not always been the primary function of the nation state (early state systems were largely about securing the absolute power of a sovereign over their territory). Crucially, however, the idea of governing has become the organizing principle behind which nation states have emerged and gained support and legitimacy. Such forms of governing can been seen in fields as diverse as the improvement of public health, the effective circulation and storage of agricultural production and provision of modern water supply systems. When connected to the practices of governing, the nation state is best thought of not so much as a territorial entity, or set of centralized ministries, but as a father figure or shepherd, constantly supervising and monitoring the family/flock to ensure needs are being met. Crucially, in the context of this chapter, Foucault recognizes that the goal of government means that nation states must carefully regulate the relationship between people and the environment, so as to ensure that the welfare of the population is not compromised in the long-term. Foucault (2007 [2004]: 96) thus states:

> The things that government must be concerned about . . . are men [sic] in their relationships, bonds, and complex involvements with things like wealth, resources, means of subsistence, and, of course, the territory, with its borders, qualities, climate, dryness, fertility, and so on.

7.2.2 States, government and the environment

Having established the relationship between nation states and sovereignty, centralization, territory and the practices of government, I want to move on to briefly consider the connections that exist between the formation of states and environmental affairs. In exploring these relations, I want to illustrate the long historical connections that exist between nation states and the transformation and management of the natural world. If we start by looking at the origins of the very first large-scale political communities, we discover that issues of environmental management were central to their formation. In his book *Oriental Despotism; A Comparative Study of Total Power*, the German-American writer Karl Wittfogel (1957) argues that the emergence of early civilizations in places such as Egypt and Mesopotamia was based on the control and management of water resources. According to Wittfogel, it was the management of

water in arid areas, in the forms of flood defences, irrigation systems and domestic supply, that necessitated the formation of large-scale political bureaucracies (for more on contemporary water government issues see Chapter 2 in this volume). Given the limited sources of available water in such areas, and the need to distribute water over large tracts of land, effective water management could not be achieved at a local level: it required the formation of what Wittfogel describes as *hydraulic empires*. While the forms of political society described by Wittfogel are very different from modern nation states, and significantly predate the formation of the first nation states, they provide an important insight into the historical connections between government and the environment. The ability to coordinate and control the supply and distribution of water over large geographical scales has continued to be a key goal of the modern state systems (see Swyngedouw, 2007; Linton, 2010). However, if securing the effective supply and distribution of water was a key factor in the establishment of political communities that occupied large tracts of space, the territorial and bureaucratic forms of nation states had other implications for human environmental relations.

With the establishment of sovereignty over significant expanses of territory, early nation state systems interpreted the exploitation of natural resources as a key factor in obtaining economic and military advantages over competing sovereignties. But the wide territorial distribution of natural resources (including agricultural land, forests and minerals) made it difficult for nation states to effectively monitor and control their environmental assets. In his account of the formation of early forms of state-sponsored scientific forestry in eighteenth-century Prussia and Saxony, James Scott describes the processes in and through which states attempted to more effectively govern the use of natural resources (Scott, 1998). According to Scott, it was crucial for early state systems to have a clear understanding of their available forest resources. During the eighteenth century, forest resources were a vital source of fuel and building material (used both on the construction of houses and ships) (see Chapter 5, this volume). Beyond this, however, forests also represented an important source of fiscal wealth: a product that could be taxed and traded in order to consolidate a nation's wealth (Scott, 1998: 12). Scientific forestry essentially enabled state systems to effectively centralize knowledge about forest resources and to predict likely shortfalls and overproductions of timber. In order for this to happen, Scott describes how scientific forestry sought to effectively simplify forests and make them more legible to state bureaucrats. This simplification process was in part based upon the planting of regimented forests, where trees were planted in geometric patterns. These planting systems made it much simpler to effectively calculate the volumetric timber yield of any government forest. This process also involved ignoring the broader socio-ecological form of woodlands. Scott (1998: 12) observes:

> Lurking behind the number indicating revenue yield were not so much forests as commercial wood, representing so many thousands of board feet of saleable timber and so many cords of firewood fetching a certain price. Missing, of course, were all of those trees, bushes and plants holding little or no potential for state revenue.

Essentially Scott describes how the establishment of nation states went hand-in-hand with the simplification and standardization of the natural world. These processes of standardization and simplification did not apply only to forests but also to water resources, agricultural land and mineral deposits. Governments consolidated their territorial power by developing ever more sophisticated maps of their environmental resources. These maps and surveys increasingly deployed standardized volumetric measures in and through which nature could be subject to government calculation (see Braun, 2002; Whitehead et al, 2007: 86–116). It is in this context that it is possible to see how the formation of

nation states not only enabled environmental resources to be utilized and exploited in different ways, but also changed the way in which the environment was seen and understood.

7.3 THINKING ABOUT STATE–ENVIRONMENT RELATIONS: GREEN ARBITERS AND ECOLOGICAL LEVIATHANS

Having established the historical connections that exist between the emergence of nation states and changing human understandings and treatments of the environment, this section briefly considers available theories that can assist in the interpretation of state–environment relations.

7.3.1 Anarchism and the ecological good life

One of the earliest groups of theories that attempted to analyse the relationship between nation states and the environment was anarchism. Anarchism is an interconnected set of political philosophies that emerged during the nineteenth century (although its basic premises are in many ways as old as philosophy itself). As a movement that rose to prominence during the nineteenth century, anarchism emerged at a time when nation states were growing in power and significance throughout the world. The word anarchism essentially translates as 'the absence of leader' (Whitehead et al, 2007: 28). Anarchism is perhaps best interpreted as a movement that seeks to oppose the formation hierarchies of power, such as those found within organized religion, science and the nation state. Anarchists argue that the formation of large systems of social organization tend to diminish the liberty and freedom enjoyed by individuals and result in the loss of creative opportunities within human life (see Bakunin, 1990; Kropotkin, 1974). In regard to human–environment relations, anarchists argue that the formation of modern states has resulted in three processes: 1) the environmental dispossession of people who effectively lose access to once commonly shared resources that are now controlled by the state (including woodlands and water courses); 2) a general decline in the ecological knowledge and capacities of people whose previous engagements with the natural world (perhaps through forest agricultural, coppicing or water management) are now administered by state officials and experts; and 3) the general alienation of people from the natural world. Prominent anarchist writers, such as Henry David Thoreau (1995), have argued that real freedom, and associated forms of personal independence, can only be achieved if individuals are able to connect more directly to the environmental systems upon which all life depends. Neo-anarchist thinkers, such as Murray Bookchin, argue that it is only through the abolition of nation states and the formation of smaller scale municipal forms of government that the exploitative treatment of nature can be redressed (Bookchin, 1986). In his book *Post-scarcity Anarchism*, Bookchin (1986) observes how hierarchical systems of social organization, such as the nation state, have consolidated humanity's collective sense of power over the natural world. Echoing the work of James Scott (1998), Bookchin also argues that the rise of hierarchical societies has been associated with nature being understood in simplified and objectified ways. As nature is increasingly removed from people's everyday experiences it becomes more and more difficult to discern how it is being used and exploited in order to support industrial society. To these ends, anarchist thought and philosophy suggest a potential paradox within the Anthropocene: namely that although humans may be responsible for more and more of what goes on in the global environment, their actual awareness and sense of responsibility for such actions may be diminishing (see Chapter 8 for a discussion of the psychological processes associated with this situation).

7.3.2 Ecological leviathans and green referees

Other theories of the modern state suggest that it may have a more positive influence on human–environment relations. Such theories tend to see states and governments as having a degree of autonomy from the political and economic interest groups that are present in any society. The state is thus seen to play a crucial role in protecting the collective interests of society, which may include the sustainable use of natural resources or the preservation of clean air and water. These are parts of the environment that are typically seen as common goods that would be neglected if individuals were allowed to simply follow their own selfish interests (see Johnston, 1996: 131–132). While such visions of the state tend to be based upon rather negative understandings of human nature (centred on a selfish gene that drives a inherently self-serving and competitive human nature, see Rifkin, 2009), they do reveal the important role that states can play in governing so-called *environmental externalities*. In economics, externalities are the unwanted side effects of economic transactions that are visited upon peoples or groups that have no part in the original activity. Like the innocent bystander in a comedy sketch who is struck by a custard pie that is intended for someone else, the problem with environmental externalities is that they do not tend to incur a cost on the actual perpetrators of the original action. In many ways states are like social custodians, to whom we collectively grant certain powers (such as law-making, policing, regulating) so that others do not unfairly visit externalities upon us. It is in this context that states are able to regulate environmental externalities such as air and water pollution, to ensure that those who economically benefit from causing environmental harm are held to account.

7.3.3 The capitalist state and the environment

A third set of theories that connect the state with environmental affairs has emerged out of work within Marxist political economy (for more on Marxism see Chapter 2 in this volume). These theories tend to depict the state as having a far from neutral role in the management of

Box 7.1 Henry David Thoreau

Henry David Thoreau was an American writer who was central to the emergence of the environmental movement in the US. While Thoreau may not have called himself an anarchist, his work and action serve as examples of the nature of anarchism and its connections to environmental philosophy. Thoreau once spent time in a Concord jail as a consequence of his unwillingness to pay taxes that were being levied in support of America's war in Mexico (Walls, 2001). It is claimed that Thoreau never paid taxes throughout his life, and this reflects his fierce sense of independence and his opposition to the controlling influence of political authorities. It is within his nature writings, however, that we discover the connections that exist between Thoreau's anarchist politics and environmental ethics. In 1845 Thoreau built a log cabin on land adjacent to Walden Pond (near Concord in Massachusetts). He then lived, in relative isolation, in this cabin for over two

Plate 7.2 Henry David Thoreau
Source: Wikimedia Creative Commons

years. During this time he compiled one of the great pieces of nature writing, *Walden: Or, Life in the Woods* (Thoreau, 1995). In *Walden*, Thoreau provides a detailed account of the impacts that his sustained immersion in the natural world had on his sense of self and identity. Thoreau thus observed, 'Not till we are lost, in other words, not till we have lost the world, do we begin to find ourselves, and realize where we are and the infinite extent of our relations' (Thoreau, 1995: 8). In classic anarchist terms, Thoreau's experience at Walden Pond, and subsequently in more remote corners of the North American wilderness, provided him with a new appreciation of nature and humanity's place within it. Thoreau was concerned that opportunities for such environmental experiences were being eroded within modern industrial and political life. Thoreau's writings were crucial to the campaigns that emerged in nineteenth-century America for the preservation of nature and that ultimately led to the establishment of the first national parks.

Key reading

Walls, L.D. (2001) 'David Henry Thoreau, 1817–62' in Palmer, J.A. (ed.) *Fifty Key Thinkers on the Environment*, Routledge, London: 106–113

socio-environmental affairs. Marxist inspired analyses of the state often depict governments as active accomplices within the capitalist exploitation of the natural world (Burkett, 1999). In order to understand this position, we need to know more about Marxist understandings of the state.

Classical Marxist accounts of the state claim that modern states have emerged primarily to support the continued accumulation of wealth, and to resolve the socio-economic and environmental problems (Marxists would say contradictions or crises, see Merrifield, 2002) that these

Box 7.2 Karl Marx

Karl Marx is one of the most influential philosophers of the nineteenth century. His ideas not only transformed philosophical thought but also lay the foundations for the formation of socialist state systems (such as the Soviet Union) in the twentieth century. Marx's most influential works were the *Communist Manifesto* (which he co-authored with Friedrich Engels in 1848) and his magnum opus, *Das Capital: A Critique of Political Economy* (the first volume of which was published in 1867). Marx's work provided a systematic critique of evolving forms of capitalist society. At the heart of Marx's critique was his analysis of the ways in which those who owned the means of production (the bourgeoisie) exploited waged labour (the working class) in order to ensure the expansion of profits. Marx's work also has important implications for how we understanding evolving social relations with the environment under capitalism. The central environmental message of Marx's work is that capitalism leads to the subordination of environmental

Plate 7.3 Karl Marx's gravestone, Highgate Cemetery, London
Source: Wikimedia Creative Commons

values to those of commercial interest (see Smith, 1984: 114). A central aspect of Marx's work concerns the impact he recognized that the formation of private property has on human–environment relations. While the formation of property enables the profit-making processes associated with capitalism to proceed, it has two impacts on socio-environmental relations. First, it enables what were once commonly shared environmental resources to be converted into commodities that can be exploited by the owners of land. Second, it tends to disconnect local communities from the land they once shared and depended upon for their survival. As these local communities increasingly sought work in the industrial sectors of expanding cities, they became alienated from the natural world with which they were once so familiar (see Smith, 1984: 116).

Key readings

Burkett, P. (1999) *Marx and Nature: A Red Green Perspective*, Macmillan, London
Harvey, D. (1996) *Justice, Nature and the Geography of Difference*, Blackwell, Oxford
Wheen, F. (1999) *Karl Marx*, Fourth Estate, London

processes of accumulation generate. It was these assumptions that famously led Marx (and his writing partner and long-term benefactor, Friedrich Engels) to proclaim that the modern state is nothing more '[t]han the committee responsible for managing the common affairs of the whole bourgeoisie' (Marx and Engels, 2004: 82) (the bourgeoisie is a term which is popular among Marxists and is used to describe the capitalist class who own the land, machinery and properties that are used in various forms of economic activity). So a Marxist tends to see the state less as a referee when it comes to environmental issues, and more as an instrument of the ruling economic class (for an excellent contemporary account of the corporate control of government see Beder, 2002; Monbiot, 2000; Frank, 2008).

Because wealth creation in capitalist societies is based upon the transformation of nature from its primary forms (wood, ores, crude oil) into tradeable commodities (furniture, aluminium foil, petroleum), Marxists argue that states tend to do two things: 1) actively enable the corporate control and transformation of the natural environment into commodities; and 2) prevent the development and/or enforcement of rules, laws and regulations that would hinder the corporate exploitation of natural resources. There are numerous examples of the state-sponsored exploitation of the environment. A much-discussed example is the historical role of the Brazilian government in facilitating the economic exploitation of the Amazonian rainforest (see Peet and Watts, 2004: 7). It is not just that Brazilian governments in the past failed to effectively protect Amazonia, but that through the use of subsidies and military power, they supported the interests of logging and ranching corporations over and above those of indigenous communities and the environment.

More recent so-called *neo-Marxist* accounts of the state diverge from classical work in important ways. Neo-Marxist state theorists, such as Jessop (1990), argue that the state is not merely the instrument of a capitalist class. Jessop argues that the state cannot simply serve the interests of an

---

Thomas Frank and the incompetent state

This video provides a commentary on Thomas Frank's fascinating 2008 book *The Wrecking Crew: How Conservatives Rule*. In this video Frank provides a description of the ways in which economic elites are not only seeking to control the apparatus of the US state, but actually seek to create a deliberately incompetent state, which is unable to govern things such as the environment competently. Search for 'Thomas Frank – The Wrecking Crew: How Conservatives Rule' on YouTube.

---

economic elite because that elite is too diverse and places very different demands on government. To try to explain this perspective in environmental terms, it is useful to consider the contemporary debate over nuclear power in the UK. The UK government has recently pledged to support the expansion of the Britain nuclear power industry in order to combat the rising prices associated with oil and gas. But in offering its support for the nuclear industry the UK government could, in the long term at least, reduce the attractiveness of investing in wind, solar and tidal energy development. Perhaps the most important insight of neo-Marxist work on the state, however, is the emphasis it places on the development of political strategies. By political strategy I am referring to the techniques that governments use to try to find a common ground between different economic interests. Returning again to the case of the energy sector, while the UK government is offering strong support to the nuclear industry, in its 2007 *Meeting the Energy Challenge: A White Paper on Energy*, the UK government justified its support for the nuclear industry alongside a 'wider energy policy' of diversification (DTI, 2007). By positioning its support for the nuclear sector alongside a wider energy policy that involves investment and support for renewable energy initiatives, the UK has developed an energy strategy that seeks to support a range of economic interests in different ways. The strategic balance forged within the UK energy white paper concerns more than merely economic interest. It also reflects a broader compromise between competing environmental priorities. Consequently, while nuclear power provides perhaps the most commercially viable approach to meeting increasing energy demands, and at the same time reducing the UK's production of greenhouse gases and supporting other renewable

energy sources serves to off-set some of the environmental concerns that surround the transportation and storage of nuclear waste.

Although theories of anarchism, eco-authoritarianism and neo-Marxism are by no means the only ways in which we can understand the connections that exist between states and the environment (for a broader overview see Whitehead et al, 2007), they do provide us with a helpful set of frameworks for interpreting why states act in certain ways when it comes to environmental affairs. The remainder of this chapter deploys these ideas of state–environment

---

Box 7.3 Beyond the state: the rise of international systems of government

While nation states continue to be prominent actors in the regulation of human environmental relations in the Anthropocene, it is important to acknowledge that environmental government capacity is increasingly being developed at international levels. In the context of the forms of large-scale environmental issues that cross national state boundaries (such as climate change, ozone depletion, acid rain, *inter alia*), it has become increasingly apparent that multilateral action is required to effectively tackle these problems. Coordinated action on the environment can be traced back to 1972 and the United Nations Conference on the Human Environment. This conference was followed by the establishment of the Montreal Protocol in 1987, which sought to eradicate the emission of CFCs (see Chapter 3 this volume). In 1992 the international community came together in Rio de Janeiro for the United Nations Conference on Environment and Development. This conference was responsible for establishing a wide range of international agreements relating to climate change, biodiversity and forest conservation.

Notwithstanding the significant number of systems that now exist for governing environments at an international scale, it has proved difficult to coordinate international action on global environmental problems. Particular tensions have emerged within climate change negotiations between certain more economically developed countries, who are keen to address the threat of a changing climate, and emerging economies (such as China, India and Brazil) that are concerned about the impact that such policies could have on their economic development. It has also proved difficult to effectively enforce international environmental agreements (such as those regulating the whaling industry) when powerful states (such as Japan) openly contest them.

Key readings

Andresen, S. et al (eds) (2000) *Science and Politics in International Environmental Regimes*, Manchester University Press, Manchester

Whitehead, M. (2006) *Spaces of Sustainability: Geographical Perspectives on the Sustainable Society*, Routledge, Abingdon

relations in order to consider two case studies. The first case study considers the London fog disaster of 1952. The second explores evolving forms of environmental government in the Florida Everglades.

7.4 GOVERNING THE AIR: THE CASE OF THE LONDON FOG DISASTER

7.4.1 Fog and social chaos

London is no stranger to the problems of air pollution. Ever since the 1307 Royal Proclamation, which forbade the burning of sea coal, state authorities have been involved in a constant struggle to regulate the quality of the air in the city (see Chapter 3). During the rapid industrialization of the city in the nineteenth century the situation deteriorated. Chemical effluents produced from the alkali industry mixed with the soot and sulphur produced by the burning of coal in industrial and domestic premises to create what came to be known as 'pea-soupers'. What is interesting about early attempts to monitor and regulate urban air pollution in the UK was that they involved both the local and national state. During the nineteenth century, many British cities formed urban corporations that, through local taxes, provided the first forms of publically provisioned sanitation and health care (this had previously been provided by private organizations and charities) (see Whitehead, 2009: 44). An important part of the work of urban corporations was to monitor and regulate atmospheric emissions. Police officers and health inspectors could thus be seen routinely traversing cities in order to monitor, and at times prosecute, atmospheric felons. Over time, sanitary authorities became responsible for enforcing national legislation on smoke nuisance and public health. Local sanitary authorities did, however, pass their own laws on air pollution and develop their own innovative ways of enforcing restrictions on atmospheric effluvia (see Mosley, 2001). To these ends, governing the British atmosphere

appears to have always involved a mix of local and national action.

Despite the various forms of national and local legislation that were enacted in order to regulate air pollution during the nineteenth and early twentieth centuries, such actions only achieved limited success. The limited success of such legislation was the result of two issues. First, it was very difficult to effectively monitor and then prosecute corporations that broke air pollution laws. In large cities it was often very difficult to attribute smoke production to any given factory. It was also very difficult for local government smoke observers to travel across rapidly expanding urban landscapes, and then to gain suitable vantage points from which to observe air pollution activities (see Whitehead, 2009: 37–66) (see Chapter 3 in this volume). Second, early laws on air pollution tended to only be applied to corporate rather than domestic premises. The reasons for this are complex, but clearly had a lot to do with the strong sense of attachment that people had to coal fires. George Orwell, for example, once suggested that coal fires were so much a part of national identity that they were the 'birthright of free-born Englishmen' (McNeill, 2000: 66).

It is precisely in this context that we start to see the challenges that exist to the state when it comes to governing human–environment relations. At one level they have to consider the needs of industry, whose atmospheric polluting activities produce wealth and employment. At another level, they must also think of the needs of the public who are subject to the various ailments that pollution produces. At yet another level, the British state had to consider the strong cultural attachments that people had to the burning of coal in the home, and the complex system of everyday practices – from cooking to washing – which had developed around coal.

The problems of governing air pollution came to a head in London in December 1952. On 5 December 1952 a thick fog settled over the city. Cold weather conditions in early December resulted in the burning of high levels of coal within households throughout the metropolis. On 5 December these high concentrations of soot and sulphur dioxide combined with the cold weather conditions to perpetuate and deepen the fog. These climatic conditions persisted for the best part of five days (with the fog only dispersing on 9 December). While, as we have already noted, London was no stranger to fog, three initial features made this pollution incident particularly problematic: 1) the unusual length of its duration; 2) the thickness of the fog, with naturally produced fog combining with emissions of smoke to produce a dense smog; 3) the geographical area that was affected by the fog: it is reported that at its peak the fog covered an area of 1000 square miles (Whitehead, 2009: 143).

These conditions conspired to generate social and economic chaos within the city of London. Transport within the city was severely affected. Aircraft had to be directed away from London; all but three of the city's bus services were cancelled (see Figure 7.4); shipping along the Thames had to be stopped; and most train services operating in the capital were cancelled (Whitehead, 2009). With public transport services so badly affected, many residents turned to motorcars. But the foggy conditions proved too much for many drivers with accident rates on the roads increasing at an alarming rate. There was, for example, an accident involving 14 vehicles in Kent, and the emergency services were stretched to breaking point by the number of calls they received in relation to road

Plate 7.4 The London fog of December 1952
Source: Getty Images

collisions. Things became further complicated on the roads because both emergency services and the Automobile Association found it extremely difficult to locate drivers who had been involved in accidents, or just broken down, in the fog.

The fog also had a detrimental affect on social life in the metropolis. Scotland Yard reported that crime rates within the city rose sharply during the fog, with higher than normal numbers of burglaries and assaults. The most serious social consequence of the fog disaster was on public health. During the fog, King Edward's Hospital Fund for London claimed that cases of respiratory illness increased fourfold, while reported cases of heart problems were three times higher than the seasonal average (Whitehead, 2009: 144). Early analyses claimed that 4000 deaths could be directly attributed to the fog disaster (see Thorsheim, 2006: 162), but this figure has recently been revised to nearer 12,000 (Whitehead, 2009).

7.4.2 The governmental response to the fog disaster

In many ways the fog disaster saw the breakdown of social and economic life in London. It was in the context of this socio-economic chaos that the British government realized that it had to take more definitive action on the air pollution problems that had plagued British cities since the industrial revolution. It did not take long for the fog disaster to be a topic of conversation within Parliament. These discussions quickly exposed some of the weaknesses in the existing systems of British atmospheric government. First, it became clear that despite the demand for reliable mortality estimates, the Department of Health was not able to produce such statistics. Furthermore, it became apparent that the government's Atmospheric Pollution Research Committee was not fit for purpose. The Committee had only met on two occasions during the whole of 1952, and did not have the capacity or expertise to provide an effective analysis of the causes and consequences of the fog disaster (Whitehead, 2009: 145). It is

important to note at this point that it is often in the face of extreme environmental problems that governments are forced to assess and reform their capacities for dealing with such problems.

In response to both the London fog disaster, and the shortcomings in atmospheric government it exposed, the British state established an expert Committee on Air Pollution in 1953 (Thorsheim, 2006). Given the serious nature of the London fog disaster, and the fact that it had affected all social classes in the city, the British state was able to gain support for drastic action on air pollution. The various recommendations of the Committee on Air Pollution led to the now famous 1956 Clean Air Act. Many claim that the 1956 Clean Air Act represents the first legislation in the world to focus explicitly on environmental pollution. What is most interesting about this piece of environmental legislation – at least in the context of the discussion presented in this chapter – is what it tells us about the role of the state within environmental governance. The 1956 Act banned the emission of dark smoke from industrial and domestic chimneys. It also created smokeless zones and smoke control areas in many cities. Finally, it established long-term provisions for the relocation of power stations away from urban centres. These actions reveal the ability of states, when freed from the influence of narrow class interest, to take fairly authoritarian action on environmental issues that can be applied in a uniform way right across a national territory. But the 1956 Clean Air Act also tells us something about the conditions that enable authoritative environmental actions by governments. By the 1950s, technologies already existed for the replacement of smoke-producing coal with cleaner gas and electricity. To these ends the 1956 Act was primarily about encouraging a form of large-scale technological substitution. The comprehensive nature of the 1956 Clean Air Act meant that it had a fairly immediate impact, and resulted in a 90 per cent decrease in sulphur emissions in London alone (O'Neill, 2000: 66).

Despite the success of the 1956 Act in addressing dark smoke pollution, it is important to note that air pollution continues to be a major problem in Britain today. In a 2007 report the Royal Commission on Environmental Pollution (a body that has now been dissolved as part of the UK government's austerity measures) estimated that air pollution was a contributory factor in 24,000 deaths in the UK every year, and that air pollution cost the UK somewhere in the region of £9.1 billion a year, through healthcare costs and work absences (Royal Commission on Environmental Pollution, 2007: 35–40). A large part of the contemporary air pollution problem in the UK is related to the rise of the motorcar (see Chapter 3). Ironically, clearing the air of dark smoke and soot has generated suitable conditions for the creation of photochemical smogs in British cities (as the sun's rays react with car exhaust fumes). Unlike the technological transfer that followed the 1956 Clean Air Act, it is clear that no viable alternative to the internal combustion engine can be introduced in order to render the contemporary automobile obsolete. In many rapidly expanding cities of the global south the situation is even worse. Here we find rapid industrialization (similar to that experienced by London in the nineteenth century) being combined with rapid increases in the rates of car ownership. It would appear that the ultimate lesson of the London fog disaster is that while governments, when suitably motivated, may be effective at tackling environmental problems in the short term, in the longer term economic change and development will always generate new, often more difficult environmental challenges.

7.5 RIVERS OF GRASS: THE US STATE AND THE FLORIDA EVERGLADES

The Everglades constitute a unique subtropical ecosystem. Located at the southern tip of the state of Florida (see Figure 7.1), the Everglades are at one and the same time one of the most environmentally important and severely threatened ecosystems in North America. The ecosystem that supports the Everglades starts to the south of the

city of Orlando with the Kissimmee River. This river flows into Lake Okeechobee, and it is to the south of this lake that the Everglades really begin. The Everglades have formed over many thousands of years as a result of the slow and steady overflow of water from the southern shore of Okeechobee. This overflow of water constitutes the hydrological system upon which the Everglades are based. But this is unlike any river you will have come across before. This is a river system that is only a few inches deep, but stretches to some 50–60 miles in width. This river flows at imperceptibly slow speeds (around 100 feet a day), but runs for about 100 miles until it reaches the Florida Bay. This distinctive environmental system has given rise to a complex and diverse ecological landscape. The landscape to the south of Okeechobee is one of swamps, sawgrass ridges, mangroves and tree islands (see Plate 7.5). This landscape is home to a rich diversity of animal life ranging from alligators to wading birds. A complex balance has emerged in this environment between the water level of the Everglades and the varied plants and animals that have made their home in the area. Small variations in the depth of the film of water covering the Everglades can have severe consequences for the ecological balance of the area (see Box 7.4).

The Everglades provide an important case study in state environmental relations, because the US federal and the Florida state governments have played central roles in supporting the economic exploitation of this region and in championing its conservation. The story of the duplicitous role of government authorities in the Everglades begins in the late nineteenth and early twentieth centuries. It was at this point in time that the first attempts were made to start to drain and reclaim significant portions of the wetlands. Local politicians, who were influenced by agro-industrial interests, supported these early attempts at drainage. The agro-industrial sector was keen to see the reclamation of the Everglades as they recognized the potential of this area to support sugar cane production (see Hollander,

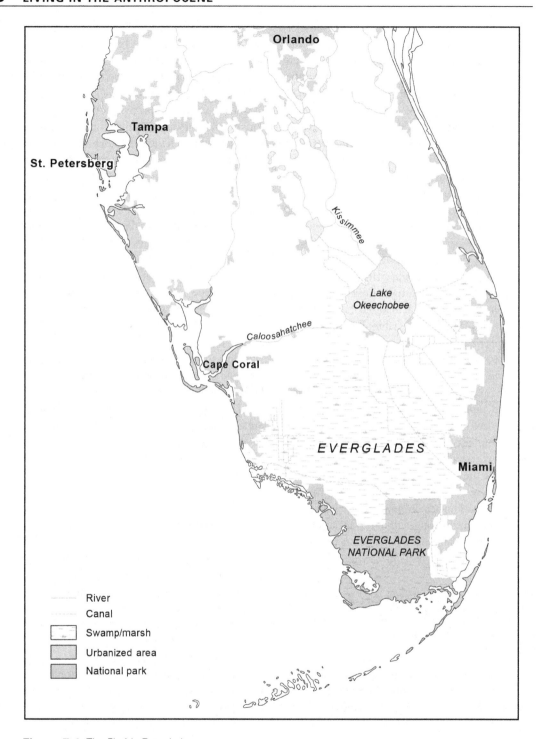

Figure 7.1 The Florida Everglades
Source: Wikimedia Creative Commons

Plate 7.5 Sawgrass prairie, Everglades National Park
Source: Wikimedia Creative Commons

Box 7.4 Apple snails and the delicate ecological balance of the Everglades

An example of the delicate ecological balance that exists in the Everglades is provided by the case of apple snails that occupy the region (see *Economist*, 2005b). Apple snails lay their eggs on the blades of grass that grow throughout the Everglades. The success of apple snails' attempts to reproduce is critically conditioned by the height at which they place their eggs on the grass stems and the level of the water in the region. If they lay their eggs too high up the grass, the stems bend and break. If the eggs are laid too low then rising waters can engulf them. Just as the survival of the snail eggs is conditioned by water levels in the Everglades, so does it determine the fortune of the other species, such as the snail kite bird, that eat the snail eggs.

Key readings

Economist (2005b) 'Water, bird, man: A vast environmental project in Florida with lessons for the post-Katrina clean-up', *The Economist*, 6 October

2008). Funding from the federal government's Internal Improvement Fund was used to support these early drainage schemes. Government policy for the Everglades shifted course in the 1920s following two major hurricanes that hit the region in 1926 and 1928. These storms resulted in severe flooding around Lake Okeechobee and the loss of many lives (Grunwald, 2006). These floods ultimately led to a new emphasis being placed on governmental authorities to provide flood defences for those living and working in the Everglades. President Herbert Hoover subsequently instructed the Army Corps of Engineers to build a new levee around Okeechobee and to construct a dam in order to control the southward flow of water from the lake into the Everglades (*Economist*, 2005b). The state-orchestrated construction of a dam around Lake Okeechobee represents a significant turning point in the nature of state intervention in the Everglades. While governmental authorities had previously sanctioned the partial drainage of sections of the Everglades in order to support the agricultural industry, by damming Lake Okeechobee, state officials were now tinkering with the hydrological systems that controlled and regulated the entire ecology of the wetlands.

The damming of Lake Okeechobee paved the way for further state-sponsored intervention within the Everglades. With the encouragement of local boosters (promoters of local areas) and sugar barons, the Army Corps of Engineers drained an area to the south of the lake to make way for large sugar plantations (see Hollander, 2008; *Economist*, 2005b). During the 1960s, the Corps of Engineers undertook a series of river diversion and dyke and levee constructions in order to allow the suburban spread of Miami and Fort Lauderdale into the Everglades (see Chapter 6 for a more detailed discussion of the political economy of the suburbanization process). Caught between the economic interests of sugar barons and urban property developers, the US state actively supported the ecological transformation and degradation of the Everglades. Estimates claim that half of the original Everglades have now been lost to farming and

urbanization (*Economist*, 2005b). But this situation is compounded by the fact that because less water is now reaching the southern end of the Everglades (due to modifications that have been made around Lake Okeechobee), salt water (and associated mangrove swamps) is moving into the Everglades at an average rate of 12 feet every year (*Economist*, 2005b).

As is typical of the complex nature of state–environment relations, national and local governments have also played an important role in trying to protect and restore the Everglades. The first attempts to try to actively protect the Everglades came in 1947, when the southern section of the wetlands was dedicated a National Park. The formation of Everglades National Park was, in part at least, a reflection of changing attitudes to the wetland system. At the heart of this change was a realization that far from being a 'stagnant swamp' in need of human reclamation, the Everglades were a dynamic riparian system in need of protection (see Stoneman, 1947). In later years, this desire for preservation was also driven by an increasing realization of the economic value of the Everglades as a tourist destination. In practical ecological terms, the Everglades National Park is ineffective at addressing the main threats to the wetlands systems. While protecting a segment of the Everglades from commercial encroachment, the National Park has no jurisdictional power to control the upstream processes that ultimately threaten the sustainability of the wetlands (namely the flow of water from Lake Okeechobee).

During the second half of the twentieth century, preventing the continued destruction of the Everglades became a primary concern of American environmental movements. These environmental interest groups were central to the formation of the Comprehensive Everglades Restoration Plan (CERP). Unlike the Everglades National Park, CERP has been designed to tackle the root causes of ecological instability in the Everglades. Through a series of 68 projects, implemented over a 30-year period, CERP will attempt to re-flood large parts of the Everglades by allowing more water to enter the wetlands. CERP was signed into law by President Bill Clinton in 2000, with an estimated $7.8 billion due to be spent during the lifetime of the project (*Economist*, 2005b). The implementation of CERP has, however, encountered a series of setbacks and difficulties. In many ways, these difficulties serve to illustrate the challenges that are associated with governmental interventions in environmental systems. The main problem is that simply allowing the natural hydrological system to return to the Everglades would have flooded large areas of productive agricultural land and generated heightened flood risks in large cities such as Miami and Fort Lauderdale. In the wake of the events of Hurricane Katrina, such flooding risks are clearly unacceptable. What CERP must thus attempt to achieve is a complex system of managed flooding, in and through which flooding in some areas of the Everglades is counterbalanced by flood control and careful water channelling in others. In this context, CERP is less a restoration project and more a watershed management plan.

It is in the context of such hydrological complexities that CERP has encountered resistance and threatened to overshoot its allocated budget. Every new engineering project has to meet the demands of environmentalists who are concerned that current re-flooding plans will not restore the Everglades, as well as those of urban residents and agriculturalists who want less flooding and greater protection. And so it is that in the swampy landscapes of the Everglades we find national and local governments metaphorically and literally up to their knees in water. In many ways, the case of the Everglades serves to illustrate that neither Marxist theories of the state (where the state serves narrowly defined class interests) nor visions of the state as a neutral referee (policing competing economic and environmental interests) effectively capture the nature of the relationships between government systems and the environment. In the world of real-time environmental management it appears that the state finds itself trying to balance different socio-economic needs

on a case-by-case basis, with some interests winning out on certain occasions, and other interests prevailing elsewhere. The latest developments in the Everglades have seen the State of Florida agreeing to buy US Sugar's operations in the region. By gradually winding-up US Sugar's operations in the Everglades, the State of Florida hopes to free up some 760km^2 of the wetland for restoration. At an estimated cost of $1.75 billion, this is an expensive way to pursue environmental restoration (Cave, 2008). It also appears to represent an intriguing case of the public sector purchasing premium corporate real estate that was only reclaimed through significant public funding in the first place.

News on the latest developments in the work of the Comprehensive Everglades Restoration Plan can be found at:

http://www.evergladesplan.org/

7.6 CONCLUSIONS

In this chapter we have explored the nature of state systems and the practices of government with which they are associated. Although the power of nation states is being partially eroded by the processes of globalization (see Chapter 5), it is clear that they represent key actors in the evolving systems of environmental management that are a part of the Anthropocene. This chapter has illustrated how the territorial reach, institutional capacities and expertise of governments makes them powerful actors in the regulation of human–environment relations. We have also seen that while various theories of the role of the state within environmental affairs exist (including anarchism, Marxism and eco-authoritarianism), they are often unable to deal with the complex nature of state–environment interactions. Through the examples of the London fog disaster and ecological restoration in the Florida Everglades, we have seen that it is impossible to characterize the state simply as environmental pariah or ecological guardian. Governments clearly reflect important frameworks in and through which humans collectively relate and respond to the environment. As we move further into the Anthropocene, however, it appears likely that states will play a vital role in addressing some of this era's greatest environmental challenges, while through other policies the very same governments will generate new ecological threats and challenges.

KEY READINGS

Hollander, G.M. (2008) *Raising Cane in the 'Glades: The Global Sugar Trade and the Transformation of Florida*, The University of Chicago Press, Chicago. This book provides a comprehensive account of the emerging role of state authorities in the Florida Everglades.

Whitehead, M. (2009) *State, Science and the Skies: Governmentalities of the British Atmosphere*, Wiley-Blackwell, Oxford. Consult this volume for more information on the London fog disaster and the broader history of atmospheric government in the UK.

Whitehead, M., Jones, R. and Jones, M. (2007) *The Nature of the State: Excavating the Political Ecologies of the Modern State*, Oxford University Press, Oxford. This volume provides a detailed introduction to state theory and its application to environmental concerns. It also introduces a series of contemporary and historical case studies of state–environment relations.

CHAPTER EIGHT

Greening the brain

Understanding and changing human behaviour

8.1 INTRODUCTION: HUMAN PSYCHOLOGY IN THE ANTHROPOCENE

In early 2009 experts from the Centre for Research on Environmental Decisions gathered at Columbia University to discuss why the human brain is not green in its orientation (Gertner, 2009). This meeting, which drew together social scientists, psychologists and economists, reflects a broader search for more sophisticated ways of understanding humans' environmental behaviours. Those gathered at Columbia University appeared to agree that there were identifiable, and significant, barriers to developing more environmentally sustainable patterns of human behaviour. At one level, experts argued that individuals tend to receive very poor feedback on the consequences of their environmental actions. If we focus on domestic energy consumption (most people's single biggest source of greenhouse gas emissions) as an example, we can identify why we often remain blissfully unaware of the environmental consequences of everyday behaviours. Within the home, energy use, and its environmental side effects, have an intangible quality. The ease with which we can switch a light on, boot-up a computer or reset a thermostat means that we often remain blissfully unaware of how much energy we are actually consuming as it makes its invisible way around our home (Jackson, 2005).

While most homes have an electricity meter, which keeps a record of energy consumption practices, these devices are usually tucked away in dusty cupboards underneath stairs and are rarely consulted. But even if we were made constantly aware of our domestic energy consumption levels (and, more importantly, what these levels meant for greenhouse gas production), the worst environmental consequences of our actions would still remain hidden from us. If we take climate change as an example, the fact that contemporary patterns of greenhouse gas production tend to only change climates over relatively long time periods, and often in distant places, means that the feedback we receive on the consequences of our action is further restricted.

Feedback is an important prompt to rational human responses to the consequences of our behaviour: we make a mistake, we observe the consequences of that mistake and we correct our behaviour to ensure that we don't make that mistake again. At the Centre for Research on Environmental Decisions' gathering at Columbia University, however, other, more emotional, prompts to human behaviours were discussed. It appears that while we often like to think of ourselves as rational decision-makers (carefully weighing up our options in order to guide our conduct), a significant portion of human behaviour is driven by more irrational prompts to action (see Damasio, 1995). We discuss the nature and significance of *more-than-irrational* human behaviours in section 8.3, but at present it is important to note that many environmental problems do not effectively activate the irrational prompts to human behaviour. Returning to the example of climate change, it appears that many of the worst environmental effects of climate change will be experienced by those living in the poorer parts of the world, who have actually contributed least to global warming. On these terms, it is argued that without a direct, emotional connection to the consequences of climate change, it will remain difficult to motivate people in more economically developed countries (who are the main producers of greenhouse gases) to reform their carbon-producing behaviours.

This chapter claims that in order to understand the types of global-scale environmental changes that are associated with the Anthropocene it is necessary to take a closer look at the psychological processes associated with human decision-making. This chapter begins by exploring the ways in which changing patterns of human behaviour (particularly in relation to consumption practices and lifestyle choices) are actively contributing to the environmental transformations that have become synonymous with the Anthropocene. The second section explores the different theories that have been developed in order to explain the ways in which humans behave in relation to the environment. Section three moves on to consider the policy initiatives and programmes that have been developed in order to try to promote more environmentally sustainable behaviour, and the assumptions that these policies make about the nature of human decision-making. The final section of this chapter focuses on new forms of environmental policy, which are drawing on the insights of behavioural psychology and neuroscience, and recognize the often-irrational basis of human behaviour when it comes to the environment. Throughout this chapter attention is drawn to the ways in which our, often overlooked, everyday routines, practices and domestic habits connect us to the Anthropocene in significant ways.

8.2 CHANGING PATTERNS OF HUMAN BEHAVIOUR AND THEIR ENVIRONMENTAL CONSEQUENCES: FORDISM AND THE GREAT ACCELERATION

In a recent article exploring the historical origins of the Anthropocene, Steffen et al (2007) introduce the notion of the 'Great Acceleration'. According to Steffen et al, the Great Acceleration represents a key stage in the history of human–environment relations, during which we see a significant increase in the rates of human-induced environmental change. According to Steffen et al, the Great Acceleration started after the end of World War II and continues to this day. While the industrial era (1800–1945) laid the foundations for the Anthropocene, Steffen et al assert that it is only in the second half of the twentieth century that the human transformation of the environment became truly global in scale. The Great Acceleration can be observed in relation to rising levels of carbon dioxide in the atmosphere, which has increased from 310ppm in 1950 to 400ppm today (half of the increase in anthropogenic carbon dioxide has occurred over the last 30 years) (see Chapter 3), rising levels of dissolved, inorganic nitrogen (used as agricultural fertilizers) in the seas (see Chapter 4) and increases in atmospheric sulphur dioxide concentrations (Steffen et al, 2007: 618) (see Chapter 3). What I am most concerned with in this chapter, however, are the socio-economic practices and patterns of human behaviour that have contributed to this acceleration.

Since the end of World War II, national economies, particularly in more economically developed countries, have experienced unprecedented rates of change and transformation. These changes are often described in relation to the emergence of a Fordist economic system (see Box 8.1). The adoption of Fordist economic practices resulted in significant changes in both the spheres of economic production and

Box 8.1 Fordism

Fordism is a term that is used to describe an economic system that emerged in the middle decades of the twentieth century. The term Fordism is derived from the organization and technological innovations in car manufacture that were developed by Henry Ford of the Ford Motor Company in Michigan (Harvey, 1989b: 125). At the heart of these innovations were the instigation of $5, eight-hour days (which represented a significant improvement on workers' pay and conditions), and the implementation of an automated car assembly line (Harvey, 1989b). Despite its origins in Ford's Michigan car plant, Fordism refers to a much broader set of socio-economic changes. In many ways, the idea of Fordism informed the welfare state systems that were constructed following the end of World War II. Welfare states did two important things: 1) they offered support for working class populations in the form of health care, improved pension schemes and unemployment benefit payments; and 2) they supported economic growth by sponsoring the development of large-scale state projects, such as road and house building, the construction of new dams and reservoirs, and the building of new airports. Such initiatives provided active state support for the mass-production/mass-consumption society. Ultimately, welfare states enabled the principles of Fordism to be applied at a national level.

Plate 8.1 The Ford building – Jericho Turnpike, Mineola, Garden City
Source: Wikimedia Creative Commons, Boston Public Library

Key reading

Harvey, D. (1989b) *The Condition of Postmodernity*, Blackwell, Oxford

consumption. In production terms, Fordism is associated with the rise of mass production and conveyer-belt technologies, which have made it possible to produce commodities in ever increasing quantities and at ever-lower cost (Harvey, 1989b: 125–129). In terms of consumption, Fordist economies saw rises in real wages and family incomes and the birth of the mass-consumption era (Harvey, 1989b: 139). In the US, for example, real wages and family incomes rose from under $15,000 (in 1986 value US dollars) in the late 1940s to $29,000 in 1986 (Harvey, 1989b: 131). In essence, Fordist societies recognized that it was only through the provision of reasonable incomes for workers that a mass market could be produced that would be able to consume the goods that were flowing off conveyer belts with such efficiency.

Fordism, and the new economies that it supported, created significant socio-economic changes. These changes can be seen at one level in rising patterns of home ownership and the emergence of a mass market for domestic appliances such as refrigerators, televisions and toasters. Rising levels of income also led to an increasing emphasis being placed on comfort and convenience within the home (Shove, 2003), with the installation of central heating systems, showers and time-saving appliances such as the washing machine and, more latterly, dishwasher. The modern home that Fordism helped to create – with its temperature-regulating thermostats and labour-saving gadgets and gizmos – was a home that required much higher levels of energy to sustain it and thus placed a much greater demand on environmental resources (see Chapter 2). Beyond energy use in the home, however, Fordism is perhaps most synonymous with the rise of the motorcar. In many ways the rise of private car ownership under Fordism reflected the pursuit of comfort and convenience that we have just described in the modern house. The car enabled people to travel over short and long distances at times of the day that best suited them. Motor vehicles also afforded people new levels of comfort while travelling: particularly with the onset of air

conditioning and in-car entertainment. The motor-car was also important because of the effects that it had on the geographical organization of people's everyday life (see Chapter 6). Suddenly, with the travelling flexibility provided by the car people could live farther away from the busy cities in which they worked. The car was thus central to the emergence of modern suburbs, with its spacious roads and large houses (see Kunstler, 1995). But while the suburbs, with their spacious housing plots, gardens and garages, reflected, the growing aspirations of a more affluent society, they were landscapes that were much more energy intensive. The low-density spatial forms of the suburbs not only meant that people would have to use more fuel to travel to work, but also to complete their everyday chores, such as visiting the doctors, collecting the children from school, and let us not forget, shopping.

Many commentators have argued that the economic downturn that afflicted many western economies in the early 1970s embodied the end of the Fordist era (Harvey, 1989b). After this point, the value of real incomes received by workers has stagnated as companies have attempted to secure their profit margins by keeping wage costs down. Two interconnected processes have, however, acted to ensure that the mass consumption practices of the Fordist era have continued up until the present day. First, the emergence of freely available credit (whether that be in the form of credit cards or product repayment schemes) has meant that while workers may have less disposable income, they can still continue to consume at high levels. Second, has been the emergence of a large and sophisticated advertising industry, which has sought to ensure that people remain keen consumers. The role and impact of the advertising industry in supporting mass consumption should not be underestimated. According to James (2007: 16), in the affluent world consumers do not need to consume 40 per cent of all the things that are produced. Our households are already replete with the things that we need to get by, and more luxury items beyond that. The advertisement industry is important because through its clever use of marketing it has been able to shift consumption practices from the realm of need to the realm of want (James, 2007).

There are certain things that we need to consume on a regular basis, such as food to eat and fuel to heat our homes. Beyond that, there are also more luxury items that we now perceive as necessary items: such as computers and televisions. Wants are different to needs to the extent that they are things that we could actually survive without. We do, however, increasingly buy things that we want because we associate the purchase of such goods and services with pleasurable feelings. Advertisements exploit our wants on a series of levels. First, there is the promotion of luxury goods and services that are sold to us on the basis that they will make our lives easier (goods in this group include leaf-blowing machines, which make tidying your garden less labour intensive, and ebook readers that remove the hassle of holding a book). Second, are goods that are associated with new experiences of fun and enjoyment (including PlayStations, mobile phones, iPods and flat screen TVs with surround-sound systems). Third, are a series of products that we may already own, but through upgrades are now presented to us as new and improved (consider the ever-shrinking size of iPods, the emergence of smart-phone technologies and the ever-changing fashions associated with

---

Box 8.2 Fordism as dystopia: Huxley's *Brave New World*

An admittedly extreme example of the potential socio-environmental problems of Fordism is provided by Aldous Huxley's famous novel *Brave New World*. *Brave New World* was first published in 1932 following Huxley's travels around America. The dystopic novel is set in AF 632, that is 632 years after the birth of Henry Ford in 1863 (or on our calendar the year AD 2495). The purpose of Huxley's novel was to explore the potential future that lay in wait for a society that was dedicated to mass production and consumption. This is why the calendar in Huxley's *Brave New World* is oriented not towards a conventional god, but to Henry Ford, who was the 'presiding deity' of this future society (Bradshaw, 1993). Huxley's Fordist future is a place of nightmares where drugs are actively used to ensure the population remains obedient and dedicated to the task of continued mass consumption. The reuse and repair of old goods is actively discouraged as a barrier to consumption (with things such as knitting and sewing becoming akin to crimes). Long-term human attachments are actively discouraged in this Brave New World with people's primary orientation being geared towards material objects. People are only valued to the extent to which they can actively contribute to the processes of mass production and consumption. Government sponsored euthanasia is commonly practised as a way of dealing with the surplus population. In essence, Huxley's novel illustrates the way in which the economic pursuit of wealth production and efficiency can alienate people from the things that we hold most dear (such as marriage, recreation, engaging with the natural world and love). While we are clearly a long way from AF 632 in the world in which we live today, it is clear that human behaviour is becoming more orientated towards the production and possession of more and more material goods, to the detriment of other human value systems.

Key readings

Huxley, A. (1994) *Brave New World*, Flamingo, London
Toffler, A. (1971) *Future Shock*, Pan Books, London

clothes and shoes). It is arguably this last category of goods (namely the ones that we already own but may wish to upgrade) that is the most important to producers and advertisers alike. While there are clear limits and heavy costs associated with the development of totally new products, the ability to refine and update existing products has been central to the continued growth of the modern consumer market. In his influential book *Affluenza*, Oliver James (2007: 16) quotes an ad executive who summed up this perspective in the following words, '"What makes this country great is the creation of want and desires, the creation of dissatisfaction with the old and outmoded"'.

The modern tendency towards the over-consumption of goods has had social and environmental consequences. At a social level, Oliver James (2007) describes the emergence of what he terms an 'affluenza virus' within the western world. Affluenza is a form of social illness, in and through which we tend to judge ourselves, and those around us, on the basis of what we have

Box 8.3 The Voluntary Simplicity Movement and the problems of overconsumption

The Voluntary Simplicity Movement (VSM) is a new social movement that emphasizes the value of simplifying our complex and increasingly stressful lives. VSM started in Seattle in the 1980s, as a reaction against the materialist culture that was emerging in Reagan-era America (Blumenthal and Mosteller, 2008). The basic message behind the movement is that people should free their lives up from the many possessions that they have accumulated, and which fill up all of our cupboards and loft spaces. Mary Grigsby, who wrote a book on VSM entitled *Buying Time and Getting By* (2004), explains the principles of the movement in the following way, 'The idea in the movement was "everything you own owns you". You have to care for it, store it. It becomes an appendage' (Blumenthal and Mosteller, 2008). On these terms, voluntary simplicity encourages individuals to avoid the burdens associated with overconsumption and re-evaluate the things that are of real importance in their lives. Thus it is clear that VSM is a social response to Fordist-era norms of economic growth and development.

Over the last few years there is evidence that VSM appears to be on the rise again (see Alexander, 2013). While the re-emergence of VSM is, in part, an attempt to encourage the development of new forms of social values, it also has practical benefits. Recent years in North America and Europe have seen the emergence of a new industry, the self-storage sector. For a fee, self-storage businesses lease storage space to households who simply have no further room in their houses and outbuildings for all the possessions they own. In many ways the self-storage sector is a logical economic outcome of Fordist-era consumption patterns.

Key readings

Alexander, S. (2013) 'Voluntary Simplicity and the social reconstruction of law: Degrowth from the grassroots up', *Environmental Values* 22: 287–308

Blumenthal, R. and Mosteller, R. (2008) 'Voluntary Simplicity Movement re-emerges', *New York Times*, 18 May

Grigsby, M. (2004) *Buying Time and Getting By: The Voluntary Simplicity Movement*, SUNY Press, Albany, NY

For more information on contemporary manifestations of VSM go to the Simplicity Institute's homepage at: http://simplicityinstitute.org/

and own, and pursue happiness through the continued purchase of more and more goods. According to James, the social and psychological pressures associated with affluenza often mean that it can be the source of significant stress and anxiety for people living in more economically developed countries. In environmental terms, our contemporary penchant for overconsumption is placing ever-greater strain on the environmental resources that are needed to produce these goods. At the same time, the rapid rate at which products are now deemed to be obsolescent (either due to technological advances or because of changing social fashions) means that society is producing ever greater amounts of waste products that must be disposed of (Toffler, 1971).

For more information on the idea of affluenza and video content of Oliver James discussing the concept go to:

http://www.selfishcapitalist.com/affluenza.html

The cultures of mass production and mass consumption mean that many residents living in affluent countries are placing unsustainable demands on the Earth's environmental resources. In a recent report on ecological footprints, the World Wildlife Fund estimated that if the whole world consumed at rates similar to that in the UK we would collectively require 3.1 planets to provide all of the resources required (if Switzerland's consumption rate was mirrored throughout the world we would require 2.8 planets, but if we all operated at India's rate of consumption we would only require 0.4 planets) (Calcott and Bull, 2007: 13). As production and consumption rates are now beginning to increase in less economically developed countries, it is clear that changing human consumption behaviours represent one of the greatest policy challenges in the Anthropocene. As we see in the section that follows, the human psychological and cultural attachment to mass-consumption lifestyles means that changing human behaviours is proving a difficult feat to achieve.

Box 8.4 Daniel Miller and alternative perspectives on the consumption society

The recent work of Daniel Miller argues that we need to develop new and more nuanced accounts of the consumer society in which we now live. Rejecting conventional accounts of mass consumption that associate it with forms of social alienation and avarice, Miller argues that there are many positive, if often overlooked, aspects to consumption. Miller attacks what he describes as the 'myth of materialism', and argues that rather than undermining authentic relations (found in friendships, family or the connections with the natural world), material possessions actually support a material culture that can facilitate a rich array of valued relations between people and things. On these terms Miller argues that more inclusive and sustainable societies can be built out of consumer practices. In order for this to happen, however, Miller argues that we need to recognize the important political role of consumption in shaping the socio-ecological relations around us.

Key reading

Miller, D. (2012) *Consumption and its Consequences*, Polity Press, Cambridge

8.3 UNDERSTANDING HUMAN BEHAVIOURS: RELIGION, SCIENCE AND IDEOLOGY

Before considering specific examples of initiatives and policies that have been developed to redirect human behaviour, it is important to outline the frameworks that have been conceived to interpret and understand human behaviours towards the environment. This section briefly outlines some of the dominant theories in and through which humanity's behaviour towards the environment has been explained. In essence, these theories provide important clues as to how affluent societies have been able to create systems of mass production and consumption that appear so indifferent to their deleterious environmental consequences.

8.3.1 The biological perspective

One group of theories concerning the nature of human behaviour towards the environment suggests that our current attitudes actually derive from the long-term evolutionary development of our collective consciousness. Evolutionary biologists operating within a Darwinian framework (see Box 8.5) argue that by the very nature of our evolution as a species – competing with other species and exploiting environmental resources as best we can – we tend to be 'by nature, aggressive, materialistic, utilitarian, and self-interested' (Rifkin, 2009: 1). On these terms, it is, perhaps, unsurprising that we remain indifferent to the environmental harm that is caused by our current lifestyles, as our very survival and development as a species may have involved an ongoing competitive struggle with the ecological systems in which we live. It is important to note that recent developments in the biological and psychological sciences suggest that we may not as a species be as self-interested and uncaring as we believed, and that we may actually be an inherently *empathic civilization* (see Rifkin, 2009 and Box 8.5).

8.3.2 Religion and its role in the ecological crisis

In a more recent historical context, some thinkers have claimed that it is the emergence of organized religion – particularly in the Judeo-Christian tradition – that lies at the centre of our exploitative environmental conducts. In a famous paper published in the journal *Science* in 1967, Lynn White Jnr argues that it was not just scientific and technological development that enabled large-scale environmental transformation but the fact that these developments often occurred in a Christian context. In his paper, *The Historical Roots of the Ecological Crisis*, Lynn White Jnr (1967) argues that Christianity (and related religions) have established a sense of separation between humans (as the chosen ones of God) and the rest of creation (the natural environment) (Nelson, 2001: 203). Lynn White Jnr claims that Christianity has generated a situation where humans believe that they are superior to the rest of nature and that human scientific and technological intervention within the environment is a moral good. On these terms, it is not so much that humans are predisposed not to care about the environment, but that we feel it is our moral duty to transform and thus improve the natural world. Lynn White Jnr recognized that many people were not overtly religious, but claimed that Christian teaching on human superiority over nature had become a commonly held belief within general society (for counter arguments on the role of religion in influencing human–environment relations, see Pepper et al, 2011). It is possible to see some of these sentiments in the voices of those who claim that the emergence of the Anthropocene necessitates ever-greater forms of human intervention into natural systems in order to ensure that they function in ways that continue to serve human designs (Crutzen, 2002).

Box 8.5 Darwin and the survival of the fittest

Charles Darwin was a naturalist who was born on 12 February 1809. By the time of his death in 1882, his scientific ideas had transformed the ways in which humans thought about themselves and their place in the world. Darwin's life was transformed by a voyage around the world that he made on the HMS *Beagle*. The observations of flora and fauna that Darwin made on this trip lay the foundations for his theory of natural selection, or the ways in which certain species thrive and prosper, while others suffer and fade. Darwin set out his ideas in a series of papers and books, the most prominent of which were *The Origin of the Species* (1859) and *The Descent of Man* (1871). At the heart of Darwin's theory of natural selection was the idea that life on Earth was essentially a struggle for existence. According to Darwin, the species, like humans, who survive and prosper do so on the basis of their ability to adapt and respond to the opportunities and constraints that are placed on them by the environments in which they live. Darwin's theories proved controversial because they suggested that humans had evolved as a species over much longer periods of geological time than were suggested in the Bible and had most probably descended from simian species.

Although Darwin's ideas had a significant impact on early geographical thought, its connections with environmental determinism (or the ways in which people and societies are shaped by their environments) has resulted in it becoming somewhat marginalized within the contemporary discipline (Castree, 2009). In relation to our discussions of the Anthropocene, Darwin's work is important because it has helped to define a powerful framework in and through which the human transformation of the natural environment has been explained and justified. In Darwinian terms, for example, it is possible to interpret the human use and exploitation of nature as part of a competitive struggle with other forms of life for the limited resources available on Earth. Darwin's theories are, however, also open to other forms of interpretation. The emphasis that Darwin places on the ability of successful species to evolve and adapt to their environment also indicates that survival is partly based upon developing intricate balances between species and the ecosystems within which they live.

Plate 8.2 An 1871 caricature of Darwin
Source: Wikimedia Creative Commons

Key readings

Darwin, C. (1998 [1859]) *The Origin of the Species*, Wordsworth Editions, Ware, Herts

Darwin, C. (2004 [1871] *The Descent of Man*, Penguin, Harlow, Essex

Box 8.6 *The Empathic Civilization*

In his book *The Empathic Civilization: The Race to Global Consciousness in a World in Crisis* the American economist Jeremy Rifkin (2009) describes an emerging paradox between *empathy* and *entropy*. On Rifkin's terms, empathy relates to the human capacity to connect with, and care for, others (both human and non-human), while entropy relates to our collective misuse of the planet's finite resources. Rifkin (2009: 2) summarizes this paradox in the following terms:

> Throughout history new energy regimes have converged with new communication revolutions, creating ever more complex societies. More technologically advanced civilizations, in turn, have brought diverse people together, heightened empathic sensitivity, and expended human consciousness. But these increasingly more complicated milieus require more extensive energy use and speed us towards resource depletion.

At the heart of Rifkin's book is an analysis of whether heightened levels of human empathy for distant people and environments can provide the basis for building a more sustainable global economy before the consequences of our misuse of the biosphere threaten the civilization we have constructed. Crucially, Rifkin draws on recent work within the study of human behaviour to argue that despite long-held beliefs that humans are inherently self-interested and individualistic, humans are a naturally empathic species. The in-built empathetic nature of humans can be seen in the empathic distress that newborn babies display when they respond to the sound of other babies' cries by crying themselves. But Rifkin (2009: 10) claims that it is the 'evolution of empathic consciousness that is the quintessential underlying story of human history', or to put things a different way, 'Empathy is the very means by which we create social life and advance civilization'.

But even if humans are essentially empathic in their nature, questions still remain about the extent to which empathy extends to the non-human world. Rifkin excitedly reflects upon the work of biologists who suggest that there are 'mirror neurons' in the brain, which mean our brains have a 'predisposition for empathetic response' when we see distress in other mammalian species (Rifkin, 2009: 14). The extent to which humans exhibit or can develop similar emotional responses to the suffering of large-scale biospheres and ecosystems remains an open question.

Key reading

Rifkin, J. (2009) *The Empathic Civilization: The Race to Global Consciousness in a World in Crisis*, Polity Press, Cambridge

8.3.3 The impact of industrial capitalism

In his influential analysis of human relations with the natural world, the geographer Neil Smith claims that it is not biology or religion but capitalism that has laid the foundations for current patterns of environmental exploitation (Smith, 1984). Smith states, 'More than any other identifi-able experience, the emergence of industrial capitalism is responsible for setting the contemporary views and visions of nature' (Smith, 1984: 1). Smith claims that modern capitalist society, with its emphasis on the free market, wealth creation and entrepreneurialism has supported the emergence of two distinct, but connected, ideologies of nature. On the one hand, Smith identifies *external ideologies of nature*, which (in keeping with the ideas of science and religion) suggest that humans are somehow separate from the rest of nature. On the other hand, Smith describes a *universal ideology of nature*, within which humans are seen to be as much a part of nature as any other component of the biosphere. While these two ideologies of nature may seem to be contradictory, Smith outlines the ways in which they have been deployed within capitalist societies to support the expanded transformation of the natural world. In relation to external ideologies of nature, Smith describes the ways in which capitalism has transformed environmental resources from parts of a complex web of ecological systems to being individual commodities that belong in the market place. This is how forests become timber, cows become beef and grass becomes fodder. Of course, by objectifying nature in this way, capitalism has made it much easier for people to accept the wholesale transformation of the natural world, as the environmental consequences of these actions become obscured by long commodity chains and complex market place exchanges. Smith also argues that universal (or internal) ideologies of nature have proved important to the capitalist enterprise. By positioning humans within a natural order of things, universal ideologies of nature have enabled ruling elites to promote a moral code of conduct, and a related sense of the virtue of labour among the working classes. While both the external and universal ideologies of nature may have originated within science and religion, Smith claims that it is only under capitalism that they became a fully functioning part of our social and economic systems, and, as such, a basis for the wholesale transformation of the global environment.

8.4 CHANGING HUMAN ENVIRONMENTAL BEHAVIOURS: BEYOND HOMOECONOMICUS

So far in this chapter we have established the changing patterns of individual human behaviour that are associated with modern forms of environ-

mental transformation, and various theories that provide explanations for these behaviours. Since the late 1960s, governments and various environmental NGOs have been striving to make human behaviours more environmentally sustainable (Bullen and Whitehead, 2005). This section of the chapter explores the types of policies and strategies that have emerged over this time period. In exploring these attempts to reform human behaviours towards the environment, this section suggests that there has been an important shift in these policy regimes. This shift has seen a move away from economically orientated policies (such as taxes and subsidies) towards more psychologically oriented initiatives (see Jones et al, 2011a; 2011b; 2013). We also see that this policy shift has been connected to a changing understanding of the nature of human behaviour.

8.4.1 Environmental policies and the rationality assumption

Throughout much of the history of modern government, bureaucrats have designed policies that assume people act in predictable and rational ways (Whitehead et al, 2011). Now, what is deemed to be a rational (or irrational) decision is hotly contested, and tends to vary between different time periods. On these terms it is best to think of rationality not in terms of the outcome of a particular decision, but in relation to how that decision was made. Understood in this way, rational action is associated with a series of interconnected practices. To act rationally involves time being taken to carefully deliberate on your best course of available action. Acting rationally also tends to rely on you being in possession of the information you need to make an informed choice and an ability to interpret this data effectively. Finally, modern understandings of rationality tend to assume that the decisions we make involve us acting in our own best interest (in other words focusing primarily on the benefits that a course of action will bring to us, and not on the wider social or environmental implications of that action).

It is a peculiar truth that the assumption that humans act in the rational ways described above does not come from the study of actual human behaviour, but from theoretical assessments of what types of behaviour are needed in order for free market economies to operate efficiently (Thaler and Sunstein, 2008). In a market economy, the efficient distribution of goods and services requires humans to respond to price signals in a rational way: buying the things that they want at the most competitive prices. Such actions are important because they indicate to entrepreneurs the products and services that people want more of, and stimulate healthy competition in the pricing of goods. On these terms, the assumption that people tend to perpetually act out of self-interest (pursuing the best prices for the goods and services that you produce and consume) is also crucial. As soon as people stop acting out of relatively narrow forms of self-interest, it becomes more likely that collusions will occur (often involving price fixing and profit manipulation), and the market place stops functioning as it should. The figure that has emerged to represent rational human action in the market within the popular conscious is *homoeconomicus* (Persky, 1995). The figure of homoeconomicus is a kind of ideal type of 'economic person' who lives out life as a rational market actor.

In the section that follows we explore some of the problems associated with the idea of homo-economicus. But for now it is important to understand why it has become such an important framework within which to understand human behaviour, and the impacts that it has had on the design of environmental policy. In the first instance, the figure of homoeconomicus has had enduring influence because a belief in rational human action suggests that you can have personal freedom (at least the economic freedom to act as you will in the market place) and social stability (in a self-regulating market system) (see Hayek, 1960; Friedman, 1982). On these terms, the market place becomes associated with the meeting of two valued human desires: 1) the desire to avoid anarchy and the break down of socio-economic systems; and 2) the preservation of personal liberty, as there is no need for an authoritarian state to administer socio-economic stability (see Chapter 7). Once the principle of rational human action is accepted it has clear policy implications. If, for example, you wish to ensure that people's actions are less damaging to the environment, there are different ways you can shape their decision-making without having to revert to the coercive powers of law and legislation (see Table 8.1). One popular strategy is to create markets within the environmental areas you wish to promote behaviour change within. In relation to climate change for example, Australia and the European Union have now established carbon markets in and through which companies can trade greenhouse gas emission permits with each other. Carbon markets work on the basis that climate change has emerged because the financial costs of greenhouse gas emissions have not previously been priced. By giving greenhouse gas emissions a price you provide a rational incentive for companies to reduce their emissions and enhance their profits (see Richardson et al, 2012: Chapter 12). Alternatively, policies could seek to ensure that individuals are receiving adequate information on the costs and consequences of their environmental actions. In this context, educational programmes and advertising campaigns can be used to ensure that information on the nature of environmental change is circulated as widely as possible. Finally, rationally oriented environmental policies also involve the use of direct financial inducements and penalties to foster more sustainable patterns of environmental behaviour. If governments want to encourage people to use less non-renewable energy resources (such as oil and gas), they can raise taxes on them (thus increasing their price), or you could offer financial subsidies that encourage people to fit renewable energy technologies in their homes (in countries such as Germany, the government offers substantial financial inducements to people wishing to fit solar panels to their homes).

Milton Friedman and being *Free To Choose*

To find out more about the rationality assumption and the principles of free market thinking on which it is based, see Milton Friedman's PBS series *Free To Choose*. Milton Friedman was a Nobel Prize-winning American economist. This eight-part series is interesting not only because it lays out the principles of free market theories, but because each episode concludes with a debate between Friedman and critics of his idea. Go to:

http://www.youtube.com/watch?v=f1Fj5tzuYBE

Table 8.1 Market-based policy options for environmental behaviour change

Policy instrument	Policy example
Creation of market	Carbon trading/pollution permits
Information provision	Green marketing/smart energy meters
Tax/financial penalty	Carbon tax/pollution fines
Subsidy	Tax breaks for micro-energy generation schemes

8.4.2 Predictably irrational environmental behaviours

Ongoing research in economics (and in particular in behavioural economics), behavioural psychology and neuroscience has started to cast serious doubts on the rationality assumptions associated with homoeconomicus (see Damasio, 1995; Le Doux, 1996; Thaler and Sunstein, 2008; Akerlof and Shiller, 2010). At the heart of these various studies has been recognition of the significant role that emotions and irrationality

play in human decision-making and behaviour. At one level, these studies have shown that in the business of our everyday lives, we rarely have the time, let alone available information, necessary to form rationalized decisions. But at a deeper level, there is increasing recognition that what we often refer to in derogatory terms as irrationality (I prefer to us the term *more-than-rational*) is actually a necessary and important part of human behaviour. More-than-rational forms of behaviours differ from rational action in important ways. First, while rational action tends to require relatively long periods of deliberation on possible actions, more-than-rational behaviours often involve on-the-spot decision-making. While we are conscious of our rational actions, more-than-rational decision-making often operates at a subconscious level (akin to an automatic response to a situation). Finally, while rationality is associated with forms of objective calculation (involving impartially weighing up available options and assessing risks), more-than-rational decision-making is often emotionally orientated (being heavily influenced by how we feel about a given situation). On these terms, irrational forms of behaviour are now routinely used as a catchall phrase to describe the emotional, automatic and subconscious aspects of action (even if in reality these modes of decision-making are often quite independent: you can be very conscious of your emotional response to a situation for example).

Irrational, or more-than-rational, decision-making is a product of two sets of processes. First, it derives from the limits, or bounds, that exist to our ability to be rational (see Simon, 1945, 1957). These limits relate to the fact that in the business of everyday life, we neither have the time nor inclination to act rationally in all of our decision-making. It is for this reason that we often use shortcuts in the decision-making process (Kahneman et al, 1982). These shortcuts, which can include following the decisions made by other people, or simply the things that we have done in the past, enable us to find a reason to act in a certain way, but not a rational basis for deciding to act. The second set of processes is more evolutionary in nature, and relates to our need to make quick, automatic, often emotion-based responses to situations (Damasio, 1995). Life is full of situations that require quick, automatic decision-making (whether it be when you are playing sport, trying to cross the road or being confronted by an aggressive person). As part of our evolutionary progression we have, of course, developed valuable decision-making structures that enable us to essentially act without thinking in such situations. The crucial insight here is to realize that not all irrational forms of decision-making are necessarily bad (Damasio, 1995; Gladwell, 2005). When used in situations where little time is available to make a judgement to act, following an instinct is an important part of human decision-making. As Malcolm Gladwell (2005) points out in his influential book *Blink: The Power of Thinking Without Thinking*, the crucial thing is to recognize when it is best to use our rational and irrational capacities to guide our behaviour.

When it comes to changing human behaviours towards the environment, simply recognizing that humans often behave in more-than-rational ways does not necessarily help those designing environmental policies. Indeed, for a long time one of the reasons that policy-makers chose to ignore the irrational dimensions of human decision-making was not because they did not realize that they were an important driver of human behaviour, but because they seemed mysterious and unexplainable (Becker, 1962). Recent research has, however, suggested that there are distinct patterns in the irrational nature of human behaviour (Thaler and Sunstein, 2008). The idea that humans are *predictably irrational* has provided an opportunity for policy-makers to begin to design environmental strategies that take into account the irrational aspects of human conduct (Ariely, 2008). These developments have been particularly important in the fields of environmental policy, where there is clear evidence that humans consistently act irrationally, with little

concern for the impacts that their short-term actions have on the long-term sustainability of the environmental systems upon which they depend.

When it comes to addressing more-than-rational human behaviours towards the environment, there is one further development that we need to be aware of. In addition to psychologically inspired developments in economic research (which have demonstrated the significant role of automatic and emotional forms of decision-making), emerging research in behavioural psychology has revealed the ways in which irrational prompts to human action can be used by policy-makers to actually change behaviours. Of course, this insight is nothing new. Corporations have for a long time been exploiting the often-subconscious aspects of our decision-making process to influence our consumption patterns (Packard, 1957; Twitchell, 1996; Frank, 1997; Cialdini, 2007). Whether it is through the use of adverts and commercials, or the clever design of supermarkets, corporations have become experts at exploiting the more-than-rational aspects of human decision-making in order to encourage us to consume their products in ever-greater quantities. The rise of commercial wants (as opposed to needs) depends not so much on rational decisions, but on the emotional suggestion that buying certain products will somehow make us feel better about ourselves (see discussion early in this chapter).

Increasingly, those concerned with environmental behaviour have recognized that it may be possible to utilize the insights of behavioural psychology to prompt human action away from mass consumption and towards more sustainable patterns of behaviour. Such strategies have the advantage that, unlike legal restrictions, or even market incentives, they are non-coercive: they do not seek to force people to act in more environmentally friendly ways, but merely to make it easier to act in such ways. Underlying such initiatives is the assumption that one of the reasons that people consistently behave in ways that harm the environment is because the world

For more on the use of psychological insights in the design of public policy go to:

http://governingtemptation.wordpress.com/.

This blog reports the findings of a three-year Leverhulme Trust research project on behaviour change. The blog draws particular attention to the ethical issues that are associated with such policies.

in which we live makes it difficult to act sustainably. We explore these ideas in greater detail in the section that follows, but for now it is important to recognize that there are some significant ethical issues associated with the development of environmental policies that draw on the insights of psychological persuasion.

8.5 POLICY PERSPECTIVES ON PRO-ENVIRONMENTAL BEHAVIOUR CHANGE

In response to emerging insights into the more-than-rational constitution of human behaviour a new set of environmental policies have emerged. These policies, which have been emerging largely over the last 10 years, are often referred to as *pro-environmental behaviour changing policies*. These new policies have drawn on the insights of behavioural economics and behavioural psychology (as well as a range of other ideas taken from fields as diverse as cognitive design and neuroeconomics, see Table 8.2). Collectively, these policies have been designed to change the behavioural norms that have emerged as part of our mass consumption societies (see Jackson, 2005) and to encourage humans to reduce the demands that they place on the biosphere. But, as we will see, the long-term effectiveness of these policies is coming under increasing levels of critical scrutiny.

Table 8.2 The intellectual influences and modes of operation associated with pro-environmental behaviour change

Intellectual influences	Modes of operation
Behavioural economics	Non-conscious priming
Behavioural psychology	Intelligent assignment
Cognitive design	Presumed consent
Engineering psychology	Mandated choices
Ethology	Anchoring
Intuitive judgement theory	Culture change
Material psychology	Channelling factors
Neuroeconomics	Collaborative filtering
Neuropsychology preference theory	Disclosure
Psychographics	Feedback
Social cognition	Self-registered control strategies
Social influence theory	Peer-to-peer pressure
Social marketing	Norm formation
Theories of affect	Choice editing
Time preference theories	Default positions
User centred design	Social marketing
Visual perception theory	

This section is split into two parts. In the first part I outline some of the environmental policies that are being developed in order to address the more-than-rational drivers of human behaviour. The second section outlines some of the critiques that have been levelled at new environmental behaviour changing policies.

8.5.1 New ways to save the planet

In their influential 2008 book, *Nudge: Improving Decisions About Health, Wealth and Happiness*, the behavioural economists Richard Thaler and Cass Sunstein outline the ways in which the emerging insights into the nature of human behaviour (outlined above) can be applied to 'saving the planet' (Thaler and Sunstein, 2008: 183–196). A central component of Thaler and Sunstein's discussion of environmental policy-making is the importance of *disclosure* (Thaler and Sunstein, 2008: 189). Disclosure simply refers to the ways in which corporations or public bodies disclose the impacts that their products and services have on the environment. A specific example of an environmental disclosure policy that Thaler and Sunstein discuss is the US government's *Toxic Release Inventory*. The Toxic Release Inventory (which was passed in 1986) requires that corporations must disclose the amount of harmful materials that they are placing in the environment. At one level, this policy may seem like a fairly traditional form of legal restriction on human behaviour. But Thaler and Sunstein point out that the Toxic Release Inventory does not require companies or individuals to necessarily stop their polluting activities. What it does do, however, is force corporations and individuals to be more aware of their relations to the environment, and how those relations may be perceived and judged by others (whether those others be consumers or peers). Interestingly, without legally requiring any form of direct change in the ways in which people behave towards the environment, the Toxic Release Inventory has been associated with significant reductions in toxic pollution

in the US (Thaler and Sunstein, 2008: 190). According to Thaler and Sunstein (2008: 191), the Toxic Release Inventory actually reflects a type of *social nudge*, or a policy, that uses people's emotional sense of how they are being judged by others as a basis for promoting pro-environmental behaviour change.

Often new environmental behaviour change policies combine aspects of disclosure with peer pressure. An interesting example of this type of initiative has been suggested in the UK. The UK government is currently encouraging energy suppliers to not only disclose to households their levels of energy consumption for the previous quarter, but how this compares to other consumers' levels of energy use (whether it be in relation to a local or national average). Using energy billing in this way attempts to achieve two main things. First, it provides a frame within which to assess your level of energy use. If you just receive your household's energy use data on your bill, it is very difficult to assess whether this is high or low, and thus in need of reform. Second, this new billing strategy generates a degree of peer-to-peer pressure, through which social norms about domestic energy use are utilized as prompts to behaviour change among higher-than-average energy-use consumers. Initiatives such as these are premised on the fact that human behaviour is influenced by a complex series of more-than-rational factors. In particular, they recognize that domestic energy use levels are not just based on individuals making rational calculations about how much energy they use, but are also influenced by long-held domestic habits and assumptions concerning what is a normal level of energy consumption.

Other initiatives have explored how it might be possible to encourage households to adopt renewable energy alternatives. Studies have shown that although household investment in solar panels, or ground heat-source technologies, can have significant long-term financial benefits to a family, only a minority of households actually invest in renewable technologies. An irrational barrier appears to exist to domestic investment in renewable technology. This barrier is often referred to as a *future bias*. Future bias exists when we prioritize present needs or desires over longer-term benefits. Future bias is clearly a major barrier when it comes to domestic investment in renewable technologies. While investment in renewable technologies involves a significant financial outlay in the present, its financial benefits are only recouped over a much longer period of time (as are its environmental benefits). The UK government's new Green Deal initiative has sought to overcome future bias in the domestic renewables sector by offering households the chance to have the up-front costs of installing new technologies met by corporations (such as the retailer B&Q or utility companies), while consumers pay this investment back with the money that they generate from reduced energy bills (see Behavioural Insights Team, 2011).

A final suite of environmental behaviour changing policies that has grown in prominence recently is those that draw on the insights of social marketing. Social marketing involves the use of commercial advertising techniques in the pursuit of public policy goals. Consequently, while corporations may use advertisements and promotions to get our attention and persuade us to buy their products, social marketing is based on the assumption that similar techniques can be used to promote more environmentally responsible behaviour. The South African energy company Eskom recently developed an interesting example of how social marketing techniques can be used to promote pro-environmental behaviours. This advert – which only lights up a small section of the billboard it is on would appear to be very ineffective during daylight hours – promotes the responsible use of energy by combining humour with the establishment of a new social norm. The billboard appeals to our collective sense of humour to the extent that it cleverly subverts the common use of excessive and unnecessary lighting in commercial roadside advertising. At the same time, by demonstrating Eskom's personal commitment to conserving energy (by

paying for a full billboard advertisement, but only using a fraction of the rented space in order to reduce energy use), the advert psychologically suggests that wasteful energy use is not normal and should always be resisted. Social marketing techniques, like those used by Eskom, are part of the new environmental behaviour changing regime to the extent that they do not use rational prompts to action, but instead appeal to the more emotional sides of our characters.

8.5.2 Critiques of environmental behaviour change policies

Environmental policies that target the more-than-rational aspects of humans are clearly based upon a more nuanced, and potentially accurate, understanding of why humans behave in harmful ways towards the environment. Notwithstanding this, these new policy initiatives have become subject to numerous critiques (see Crompton, 2010; Shove, 2010; Jones et al 2013; Whitehead et al, 2011).

At one level, ethical concerns have been raised about these new policies. Given that many of these initiatives seek to exploit people's emotional (often automatic) responses to behavioural prompts, some commentators have suggested that they may reflect forms of social manipulation that, even when pursued in the common environmental good, tend to undermine the democratic process (Whitehead et al, 2011). At another level, others have questioned whether such policies will actually be effective when it comes to reforming long-term patterns of environmental behaviour (Crompton, 2010; Jones et al, 2013: Chapter 6). In a recent report developed by a series of environmental charities and campaign groups (including the WWF, Oxfam, Campaign to Protect Rural England and the Climate Outreach and Information Network), it has been argued that contemporary behaviour changing policies are only likely to be effective over short periods of time. The *Common Cause* report argues that a lot of environmental behaviour change policy uses

FROM THE BLOG

Load profiles and changing household energy use

I recently attended a fascinating workshop on household energy practices at Durham University (hosted and organized by Harriet Bulkeley and her research team). The workshop was for a Social Science Advisory Group, which has been established to advise on Northern Powergrid and British Gas's Customer Led Network Revolution. This scheme centres around the largest smart-grid project in the UK (involving 14,000 homes and costing £54 million to implement). While our discussions were broad ranging and considered the potential impacts of smart metering and in-house energy displays on household energy usage, one of the most interesting things about the workshop for me was the perspective it provided on the processes that are driving the restructuring of the domestic and commercial energy market in the UK.

While the move towards smart-grids and meters is, of course, being driven by a desire to reduce, in aggregate, household energy use, and thus help the UK along the road to a lower carbon economy, it is also being conditioned by issues of daily household demand and energy security.

Energy use load profiles produced by the UK's national grid reveal that there are significant daily fluctuations that exist in British energy use (with the peak usage in the morning and evening periods). It is interesting to note that with the onset of the low-carbon electric economy, these peak energy use periods are likely to see more energy demands being placed upon them (as people plug in their electric cars after returning home from a long day of work). Given the great pressures that such load profiles place on energy supply networks during peak periods, energy suppliers are not only interested in how to make the home more energy efficient, but also how to redistribute energy use throughout the day.

The redistribution of energy use has, of course, been a long-term concern of energy suppliers. As a previous user of storage heaters I was able to make the most of the low, off-peak energy tariffs associated with the so-called Economy Seven initiative. But current discussions about the timing of domestic energy practices have interesting implications for behaviour change policies. It appears that shifting people's TV watching practices from the peak evening slot of 7–9 pm will be difficult, as will moving the timing of when people cook their evening meals. There may be more flexibility, however, as to precisely when people choose to take a bath/shower or put their washing machines on. New tariffs are being used to incentivize off-peak energy use, but as all UK homes join smart grids over the next decade, it will be interesting to see just how flexible our domestic energy use routines actually are.

short-term psychological strategies to shift behaviour, while not addressing our deeper socio-cultural values (see Crompton, 2010). As we discussed in the first section of this chapter, in the western world many of our values are oriented towards the goals of mass consumption and the pleasures and prestige that we attach to the ownership of goods. The *Common Cause* report

suggests that even if behaviour changing policies may nudge households to fit loft insulation, or invest in renewable energy technologies, these behaviour changes are likely to be offset by the continuing value that we place on the accumulation of wealth and possessions. The *Common Cause* report thus suggests that public policy should devote more attention to promoting the adoption of more non-materialistic, egalitarian and biocentric values among the population, as it is only a shift in these forms of value that is likely to bring a long-term transformation in the ways in which people treat the environment (see also Jackson, 2009).

8.6 CONCLUSIONS

In this chapter we have considered the patterns of behaviour that characterize human–environment relations in the Anthropocene. Drawing on Steffen et al's (2007) notions of the Great Acceleration, this chapter began by describing the ways in which human socio-economic development over the last 200 years has placed escalating demands on the biosphere and given rise to the very notion of the Anthropocene. This chapter explored various theories that have been developed to explain how and why human behaviour has developed in ways that are increasingly unsustainable. These ideas include biological theories of human evolution; accounts of the impact of modern religion on human understandings of their relation to the environment; and the ideologies of nature that have emerged in the wake of the rise of modern capitalism.

The final two sections of this chapter explored the types of policies that have been emerging over the last 40 years, which have sought to redirect the environmental conduct of humans. In these sections we explored a significant policy shift that has seen environmental policies move from a narrow focus on the rational prompts that exist to human action (whether the provision of information or economic incentives) to a concern with the more-than-rational drivers of human conducts (including fear, joy, social pressures).

Ultimately, it appears that environmental policy is now based upon a more complex understanding of the nature of human behaviour that recognizes that behaviour is driven by a mixture of biological, psychological and social forces. Ultimately, this chapter casts doubt on the long-term success of policies that target the more-than-rational drivers of human action. As many in the environmental movement have recognized for some time, it appears that only a significant reorientation of the types of human values that have emerged during the Anthropocene (which prioritize economic growth, the accumulation of possessions and associated forms of social status) is likely to cause a slowing down of the Great Acceleration that we are still a part of.

KEY READINGS

Steffen, W., Crutzen, P.J. and McNeill, J.R. (2007) 'Are humans now overwhelming the great forces of nature?', *Ambio* 36(8): 614–621. This short article provides a useful introduction to the idea of the Great Acceleration and its socio-ecological implications.

Thaler, R. and Sunstein, R. (2008) *Nudge: Improving Decisions About Health, Wealth and Happiness*, Yale University Press, London. This accessible book explains the key ideas that have informed the emergence of new forms of behaviour changing policy discussed in this chapter. It also explores in detail how these policies have been adopted within the environment sector.

Whitehead, M., Jones, R. and Pykett, J. (2011) 'Governing irrationality, or a more-than-rational government: Reflections on the re-scientization of decision-making in British public policy', *Environment and Planning A* 43(12): 2819–2837. This paper provides a critique of the ethics and efficacy of contemporary environmental behaviour change policies.

CHAPTER NINE

Conclusions

Misanthropy, adaptation and safe operating spaces

From the upper reaches of the atmosphere to the depths of underground aquifers, throughout the chapters of this volume we have explored the nature and extent of the human transformation of the global environment. This socio-ecological journey has seen us follow the flow of the carbon, nitrogen and hydrological cycles, the international trade in timber, the extraction and transportation of crude oil, and the transnational movement of atmospheric sulphur dioxide. Throughout this book we have also considered the extent to which the human transformation of the environment represents something geologically significant: the birth of a new geological epoch known as the Anthropocene. We have seen that the current rates of resource extraction; climate change; the forcing of the nitrogen cycle; soil degradation; deforestation; and urbanization all point to something significant going on in the nature of human–environment relations. Notwithstanding these observations, this volume has not sought to provide a definitive answer as to whether the current patterns of environmental change require a new geological designation. It is now expected that the Anthropocene Working Group of the Subcommission on Quaternary Stratigraphy will make its formal recommendation to the International Commission on Stratigraphy in 2016. But regardless of this ruling, it is clear that the scale and scope of the human transformation of the Earth requires that we develop more

sophisticated ways of understanding the processes that are driving environmental change. To these ends, this volume has sought to show the importance of combining scientific and social scientific understanding of contemporary patterns of environmental transformation. This volume has thus shown that studying the Anthropocene requires that we understand aspects of atmospheric chemistry alongside urbanization, soil leaching and global commodity markets, the nitrogen cycle and political theories of the state, global heat balances and the psychology of human decision-making.

In addition to establishing the importance of combining scientific and social scientific frameworks, this volume has also advocated the value of studying the Anthropocene in the context of particular geographical places. There is a tendency when speaking of the Anthropocene to talk about it in sweeping geological terms and global-level events. While such talk is, in many ways, inevitable, care must be taken to ensure that while we recognize we are all living through a significant period of environmental change, the consequences and effects of these changes are felt with varying degrees of intensity in different places. In many ways, it is an awareness of the geography of the Anthropocene that is at the heart of contemporary studies of socio-ecological vulnerability, resilience and adaptation (see Adger et al, 2005; Adger, 2006; Folke, 2006). This work considers the ability of different peoples and environments to be able to both endure harmful patterns of environmental change and to adopt new practices and processes that ensure their future wellbeing is not compromised. On these terms, it is becoming increasingly clear that it is the poorest in the world (and in particular those who rely on dryland agriculture) that are going to be most vulnerable to the consequences of climate change. However, it appears that those living in more economically developed countries will be able to invest most heavily in adaptation measures (from flood defences to new crop technology) and protect themselves from the worst effects of living in the late Anthropocene (Hodson and Marvin, 2009).

Although thinking about the Anthropocene appears to require us to place the human subject at the centre of environmental studies, there are inherent dangers within this anthropocentric perspective. At one level, studying the Anthropocene, and the tremendous environmental changes that humans have initiated, can have a slightly triumphal feel. Consequently, while we may worry over the long-term environmental impacts of human development over the last 6000 years, some may take solace in the fact that humans are now in charge of the biosphere and are better placed than ever to address the ecological problems they have created. But this immodesty can lead us to forget that humans are still only one part of the planetary system they are transforming so rapidly. Failing to acknowledge the natural limits that exist to human development could have dreadful consequences for both the health of the planet and human wellbeing. At another level, however, discussion of the Anthropocene can be a fertile breeding ground for misanthropy (the hatred of humans). Those of a deep green perspective, for example, argue that the response to the current imbalances in the global ecosystem should involve a return to a natural balance, in and through which the needs of the environment are prioritized over those of humans. But if the Anthropocene illustrates anything at all, it is the fact that we cannot go back to nature (even if we wanted to). The global environment is now an irrevocable product of society and nature. The best we can do is to find a way of living as sustainably as we can in the brave new world we have created.

In a recent article for the journal *Nature*, Rockström et al (2009) consider how we could chart a just and sustainable course through the Anthropocene. On these terms, they establish what they call *safe operating limits* for different environmental processes, including climate change, oceanic acidification, stratospheric ozone depletion, the nitrogen and phosphorous cycles, global freshwater use and biodiversity loss, *inter alia*. Rockström et al's analysis indicates that biodiversity loss, climate change and the global nitrogen cycle are already operating outside of safe limits. They also point out that failure to address these overshoot areas may lead to environmental change in other areas transgressing safe limits (the most obvious example being climate change resulting in accelerating patterns of land-use change in the form of desertification). While the thresholds identified by Rockström et al are obviously open to question, it appears that if we are to create a safe operating global system for humanity we will need to understand the nature of human–environment relations in ever more sophisticated ways.

Bibliography

Adams, W.M. (2001) *Green Development: Environment and Sustainability in the Third World*, 2nd edition, Routledge, London

Adger, N.W. (2006) 'Vulnerability', *Global Environmental Change* 16(13): 268–281

Adger, W.N., Arnell, N.W. and Tompkins, E. (2005) 'Successful adaptation to climate change across scales', *Global Environmental Change* 15: 77–86

Akerlof, G.A. and Shiller, R.J. (2010) *Animal Spirits: How Human Psychology Drives the Economy, and Why it Matters for Capitalism*, Princeton University Press, Princeton, NJ

Alexander, S. (2013) 'Voluntary simplicity and the social reconstruction of law: Degrowth from the grassroots up', *Environmental Values* 22: 287–308

Allen, J. (2003) *Lost Geographies of Power*, RGS-IBG Book Series, Blackwell, Oxford

Andresen, S., Skodvin, T., Underdal, A., Wettestad, J., et al (eds) (2000) *Science and Politics in International Environmental Regimes*, Manchester University Press, Manchester

Ariely, D. (2008) *Predictably Irrational. The Hidden Forces that Shape our Decisions*, Harper Collins, London

Bakunin, M. (1990) *Statism and Anarchy*, Cambridge University Press, Cambridge

Batterbury, S.P.J., Warren, A. and Osbahr, H. (2001) 'Soil erosion in the West African Sahel: A review and application of a "local political ecology" approach in South West Niger', *Global Environmental Change* 11: 79–96

BBC (2007) 'Palm oil warning for Indonesia', online at: http://news.bbc.co.uk/1/hi/world/asia-pacific/7084306.stm (accessed 2 February 2012)

BBC (2011) 'What is fracking and why is it controversial?', online at: http://www.bbc.co.uk/news/uk-14432401 (accessed November 2012)

BBC (2012a) 'China becomes more urban in historic population shift', online at: http://www.bbc.co.uk/news/world-asia-china-16588851 (accessed 24 February 2012)

BBC (2012b) 'Chut Wutty's memory spurs anti-logging campaign', *BBC News Magazine,* online at: http://www.bbc.co.uk/news/magazine-18032883 (accessed 15 May 2012)

Becker, G. (1962) 'Irrational action and economic theory', *Journal of Political Economy* 70: 153–168

Beder, S. (2002) *Global Spin: The Corporate Assault on Environmentalism*, Green Books, Cambridge

Behavioural Insights Team [with the Departments of Energy and Climate Change and Communities and Local Government] (2011) *Behaviour Change and Energy Use*, Cabinet Office, London

Bernstein, S. (2000) 'Ideas, social structure and the compromise of liberal environmentalism', *European Journal of International Relations* 6: 464–512

Bernstorff, A. and Stairs, K. (2000) *POPs in Africa, Hazardous Waste Trade 1980–2000. Obsolete Pesticide Stockpiles. A Greenpeace Inventory*, Greenpeace, Amsterdam

Biermann, F., Abbott, K., Andresen, S., Bäckstrand, K., Bernstein, S., Betsill, M.M., Bulkeley, H., Cashore, B., Clapp, J., Folke, C., Gupta, A., Gupta, J., Haas, P.M., Jordan, A., Kanie, N., Kluvánková-Oravská, T., Lebel, L., Liverman, D., Meadowcroft, J., Mitchell, R.B., Newell, P., Oberthür, S., Olsson, L., Pattberg, P., Sánchez-Rodríguez, R., Schroeder, H., Underdal, A., Camargo Vieira, S., Vogel, C., Young, O.R., Brock, A. and Zondervan, R. (2012) 'Navigating the Anthropocene: Improving earth system governance', *Science* 335: 1306–1307

Bilger, B. (2013) 'The Martian chronicles: A new era of planetary exploration', *The New Yorker*, 22 April: 64–89

Blaikie, P. (1985) *The Political Ecology of Soil Erosion in Developing Countries*, Longman, Harlow

Blaikie, P., Cameron, J. and Seddon, D. (1980) *Nepal in Crisis: Growth and Stagnation at the Periphery*, Clarendon Press, Oxford

Blumenthal, R. and Mosteller, R. (2008) 'Voluntary simplicity movement re-emerges', *New York Times*, 18 May

Bookchin, M. (1986) *Post-scarcity Anarchism*, Black Rose Books, Quebec

Boserup, E. (1965) *The Conditions of Agricultural Growth: The Economics of Agrarian Change under Population Pressure*, Allen & Unwin, London

Botkin, D. (1992) *Discordant Harmonies: An Ecology for the Twenty First Century*, Oxford University Press, Oxford

Bradshaw, D. (1993) 'Introduction' in Huxley, A. (1994) *Brave New World*, Flamingo, London: ix–xix

Braun, B. (2000) 'Producing vertical territory: Geology and governmentality in late-Victorian Canada', *Ecumene* 7: 7–46

Braun, B. (2002) *The Intemperate Rainforest: Nature, Culture and Power on Canada's West Coast*, University of Minnesota Press, Minneapolis

Bridge, G. (2010) 'Geographies of peak oil: The other carbon problem', *Geoforum* 41: 523–530

Bridge, G. (2012) 'Gaslands', *Area* 44: 388–390

Bridge, G. and Wood. A. (2010) 'Less is more: Specters of scarcity and the politics of resource access', *Geoforum* 41: 565–576

Brimblecombe, P. (1987) *The Big Smoke*, Methuen, London

Bryant, R. (1998) 'Power, knowledge and political ecology in the third world', *Progress in Physical Geography* 22: 79–94

Bullard, R. and Wright, B. (eds) (2009) *Race, Place and Environmental Justice After Hurricane Katrina: Struggles to Reclaim, Rebuild and Revitalize New Orleans and the Gulf Coast*, Westview Press, Boulder, CO

Bullen, A. and Whitehead, M. (2005) 'Negotiating the networks of space, time and substance: A geographical perspective on the sustainable citizen', *Citizenship Studies* 9(5): 499–516

Bulmer, M. (1984) *The Chicago School of Sociology: Institutionalization, Diversity, and the Rise of Sociological Research*, University of Chicago Press, Chicago

Burkett, P. (1999) *Marx and Nature: A Red Green Perspective*, Macmillan, London

Calcott, A. and Bull, J. (2007) *Ecological Footprint of British City Residents*, WWF, Woking

Carson, R. (1962) *Silent Spring*, Houghton-Mufflin Company, Boston, MA

Castells, M. (1983) *The City and the Grassroots: A Cross-cultural Theory of Urban Social Movements*, University of California Press, Berkley, CA

Castree, N. (2009) 'Commentary: Charles Darwin and the geographers', *Environment and Planning A* 41: 2293–2298

Cave, D. (2008) 'Florida to buy sugar maker in bid to restore Everglades', *New York Times*, 25 June

Chapman, P. (2007) *Jungle Capitalists: A Story of Globalization, Greed and Revolution*, Canongate Books, Edinburgh

Christoff, P. (2010) 'Cold climate at Copenhagen: China and the United States at COP15', *Environmental Politics* 19: 637–656

Cialdini, R.B. (2007) *Influence: The Psychology of Persuasion*, Harper Collins, New York

Coelho, S. (2009) 'British ancient forests were patchy', online at: http://planetearth.nerc.ac.uk/news/story.aspx?id=608 (accessed 28 January 2013)

Cohen, A.J., Ross Anderson, H., Ostro, B., Pandey, K.D., Krzyzanowski. M., Künzli, N., Gutschmidt, K., Pope, A., Romieu, I., Samet, J.M. and Smith, K. (2005) 'The global burden of disease due to outdoor air pollution', *Journal of Toxicology and Environmental Health A* 68: 1301–1307

Collier, P. (2010) *Plundered Planet: How to Reconcile Prosperity with Nature*, Penguin, London

Corcoran, P.B. (2001) 'Rachel Carson' in Palmer, J.A. (ed.) *Fifty Key Thinkers on the Environment*, Routledge, London: 194–200

Cresswell, T. (2006) *On The Move: Mobility in the Modern Western World*, Routledge, New York

Crompton, T. (2010) *Common Cause: The Case for Working with Our Cultural Values*, WWF-UK, London

Cronon, W. (1991) *Nature's Metropolis: Chicago and the Great West*, W.W. Norton and Company, London

Crutzen, P.J. (2002) 'Geology of mankind', *Nature* 415(3): 23

Crutzen, P.J. and Stoermer, E.F. (2000) 'The "Anthropocene"', *Global Change Newsletter* 41: 17–18

Damasio, A. (1995) *Descartes' Error: Emotion, Reason, and the Human Brain*, Picador, London

Darwin, C. (1998 [1859]) *The Origin of the Species*, Wordsworth Editions, Ware, Herts

Darwin, C. (2004 1871) *The Descent of Man*, Penguin, Harlow, Essex

Dauvergne, P. and Lister, J. (2011) *Timber*, Polity Press, Cambridge

Davis, M. (1999) *Ecology of Fear: Los Angeles and the Imagination of Disaster*, Picador, London

Davis, M. (2007) *Planet of Slums*, Verso, London

Dent, D. et al (2007) 'Land' in UNEP (ed.) *GEO4: Global Outlook – Environment and Development*, UNEP, Nairobi: 80–114

DTI (Department of Trade and Industry) (2007) *Meeting the Energy Challenge: A White Paper on Energy*, HM Government, London

Economist (2003) 'Thought control', *The Economist*, 9 January

Economist (2005a) 'The shaming of America', *The Economist*, 10 September

Economist (2005b) 'Water, bird, man', *The Economist*, 8 October

Economist (2011) 'The Anthropocene: A manmade world', *The Economist*, 26 May

Eden, S. and Bear, C. (2010) 'Third sector global environmental governance, space and science: Comparing fishery and forestry certification', *Journal of Environmental Policy & Planning* 12: 83–106

Ehrlich, P.R. (1968) *The Population Bomb*, Ballantine Books, New York

Ehrlich, P. and Ehrlich, A. (1996) *Betrayal of Science and Reason*, Island Press, California

EIA (Environmental Investigation Agency) (2007) *Attention Wal-Mart Shoppers: How Wal-Mart's Sourcing Practices Encourage Illegal Logging and Threaten Endangered Species*, Environmental Investigation Agency, London

EIA (Environmental Information Adminstration) (2011) 'Energy in brief, shale gas', Environmental Information Administration, Washington, DC

FAO (Food and Agriculture Organization of the United Nations) (1999) *Inventory of Obsolete, Unwanted and/or Banned Pesticides. Prevention and Disposal of Obsolete and Unwanted Pesticide Stocks in Africa and the Near East*, Food and Agriculture Organization of the United Nations, Rome

FAO (2004) *State of the World's Forests*, FAO, Rome

FAO (2007) *State of the World Forests*, FAO, Rome

FAO (2012) 'Land resources', online at: http://www. fao.org/nr/land/degradation/en/ (accessed 2 December 2013)

Flavin, C. (2007) 'Preface' in World Watch Institute (ed.) *State of the World, 2007: Our Urban Future*, World Watch Institute, Washington, DC

Folke, C. (2006) 'Resilience: The emergence of a perspective for social–ecological systems analysis', *Global Environmental Change* 16(3): 253–267

Foucault, M. (2007 [2004]) *Security, Territory, Population – Lectures at the Collège de France 1977–1978* in Senellart, M. (ed.) and Burchell, G. (trans.), Palgrave Macmillan, Basingstoke

Frank, T. (1997) *The Conquest of Cool: Business Culture, Counterculture, and the Rise of Hip Consumerism*, University of Chicago Press, London

Frank, T. (2008) *The Wrecking Crew: How Conservatives Rule*, Harvill Secker, London

Friedman, M. (1982) *Capitalism and Freedom*, University of Chicago Press, Chicago

Gertner, J. (2009) 'Why isn't the brain green?', *New York Times*, 19 April, online at: http://www. nytimes.com/2009/04/19/magazine/19Science-t. html?pagewanted=all&_r=0 (accessed 2 December 2013)

Ghana Business News (2009) 'Europe's e-waste in Africa', online at: http://www.ghanabusiness news.com/2009/05/09/europes-e-waste-in-africa/ (accessed 6 March 2012)

Gibson-Graham, J.K. and Roelvink, G. (2010) 'An economic ethics for the Anthropocene', *Antipode* 41: 320–346

Gillis, J. (2010) 'A scientist, his work and a climate reckoning', *New York Times*, 21 December

Girardet, H. (2006) *Creating Sustainable Cities*, Green Books, Totnes

Gladwell, M. (2005) *Blink: The Power of Thinking Without Thinking*, Penguin Books, London

Gladwell, M. (2009) *Outliers: The Story of Success*, Penguin, London

Glass, A. (2003) 'Nitrogen use efficiency of crop plants: Physiological constraints upon nitrogen absorption', *Critical Reviews in Plant Sciences* 22(5): 453–470

Gray, J. (1998) *False Dawn: The Delusions of Global Capitalism*, Granta Books, London

Grigsby, M. (2004) *Buying Time and Getting By: The Voluntary Simplicity Movement*, SUNY Press, Albany, NY

Grunwald, M. (2006) *The Swamp: The Everglades, Florida, and the Politics of Paradise*, Simon & Schuster, New York

Guardian (2008) 'Soil erosion threatens land of 100 million Chinese survey shows', *The Guardian*, 21 November

Hacking, I. (1983) *Representing and Intervening: Introductory Topics in the Philosophy of Natural Science*, Cambridge University Press, Cambridge

Hajer, M. (1995) *The Politics of Environmental Discourse: Ecological Modernization and the Policy Process*, Oxford University Press, Oxford

Hall, P. (2003) 'The sustainable city in an age of globalization' in Girard, L.F., Forte, B., Cerreta, M., De Toro, P. and Forte, F. (eds) *The Human Sustainable City: Challenges and Perspectives from the Habitat Agenda*, Aldershot, Ashgate: 55–69

Handwerk, B. (2005) 'New Orleans levee not built for worst case events', *National Geographic News*, 2 September

Hanna, K.S., Clarke, D.A. and Slocombe, D.S. (eds) (2008) *Transforming Parks and Protected Areas: Policy and Governance in a Changing World*, Routledge, New York

Haraway, D. (1991) *Simians, Cyborgs and Women: The Reinvention of Nature*, Routledge, New York

Hartmann, T. (2001) *The Last Hours of Ancient Sunlight*, Hodder, Chippenham

Harvey, D. (1973) *Social Justice and the City*, Arnold, London

Harvey, D. (1977) 'Population, resources and the ideology of science' in Peet, R. (ed.) *Radical Geography: Alternative Viewpoints on Contemporary Social Issues*, Methuen, London: 213–242

Harvey, D. (1985a) *The Urbanization of Capital*, Blackwell, Oxford

Harvey, D. (1985b) *Consciousness and the Urban Experience*, Blackwell, Oxford

Harvey, D. (1989a) *The Urban Experience*, Blackwell, Oxford

Harvey, D. (1989b) *The Condition of Postmodernity*, Blackwell, Oxford

Harvey, D. (1989c) 'From managerialism to entrepreneurialism', *Geografiska Annaler, Series B – Human Geography* 71b: 3–17

Harvey, D. (1996) *Justice, Nature and the Geography of Difference*, Blackwell, Oxford

Harvey, D. (2003) *The New Imperialism*, Oxford University Press, Oxford

Harvey, D. (2005) *A Brief History of Neoliberalism*, Oxford University Press, Oxford

Harvey, D. (2006) *Spaces of Global Capitalism: Toward a Theory of Uneven Geographical Development*, Verso Books, London

Hayek, F. A. (1960) *The Constitution of Liberty*, Routledge, London

Heinberg, R. (2007) *Peak Everything: Waking up to the Century of Decline in Earth's Resources*, Clairview Books, Forest Row

Held, D., McGrew, A.G., Goldblatt, D. and Perraton, J. (1999) *Global Transformations: Politics, Economics and Culture*, Polity Press, Cambridge

Hemmingsen, E. (2010) 'At the base of Hubbert's Peak: Grounding the debate on petroleum scarcity', *Geoforum* 41: 531–540

Heynen, N., Kaika, M. and Swyngedouw, E. (eds) (2006) *In the Nature of Cities: Urban Political Ecology and the Politics of Urban Metabolism*, Routledge, London and New York

Hodson, M. and Marvin, S. (2009) 'Urban ecological security: A new paradigm?', *International Journal of Urban and Regional Research* 33: 193–215

Hollander, G.M. (2008) *Raising Cane in the 'Glades: The Global Sugar Trade and the Transformation of Florida*, University of Chicago Press, Chicago

Homer-Dixon, T. F. (1999) *Environment, Scarcity and Violence*, Princeton University Press, Princeton, NJ

Hudson, J.C. (2011) 'Agriculture' in Wishart, D.J. (ed.) *The Encyclopedia of the Great Plains*, online at: http://plainshumanities.unl.edu/encyclopedia/ (accessed 28 January 2013)

Hulme, M. (2009) *Why We Disagree About Climate Change: Understanding Controversy, Inaction and Opportunity*, Cambridge University Press, Cambridge

Huxley, A. (1994) *Brave New World*, Flamingo, London

Independent (2012) 'Chut Wutty: Anti-logging campaigner', 27 April, online at: http://www.independent.co.uk/news/obituaries/chut-wutty-antilogging-campaigner-7685147.html (accessed 15 May 2012)

International Energy Agency (2012) *Key World Energy Statistics*, IEA, Paris

INTERPOL/World Bank (undated) *Chainsaw Project: An International Perspective On Law Enforcement in Illegal Logging*, INTERPOL/World Bank, Rome

IPCC (Intergovernmental Panel on Climate Change) (2007) Climate Change 2007: Synthesis Report. Contribution of Working Groups I, II and III to the Fourth Assessment Report of the Intergovernmental Panel on Climate Change [Core Writing Team, Pachauri, R.K and Reisinger, A. (eds)], IPCC, Geneva

Jackson, T. (2003) 'Sustainability and the "Struggle for Existence": The critical role of metaphor in society's metabolism', *Environmental Values* 13: 289–316

Jackson, T. (2005) *Motivating Sustainable Consumption: A Review of Evidence on Consumer Behaviour and Behavioural Change*, Sustainable Development Research Network, Guilford

Jackson, T. (2009) *Prosperity Without Growth: Economics for a Finite Planet*, Earthscan, London

James, O. (2007) *Affluenza: How to be Successful and Stay Sane*, Vermilion, Reading

Jessop, B. (1990) *State Theory: Putting the Capitalist State in its Place*, Polity Press, Cambridge

Jie, D. (2010) 'Chinese soil experts warn of massive threat to food security', *SciDev.Net*, 5 August

Johnston, R. J. (1996) *Nature, State and Economy: A Political Economy of the Environment*, 2nd edition, Wiley & Sons, Oxford

Jones, R., Pykett, J. and Whitehead, M. (2011a) 'Governing temptation: On the rise of libertarian paternalism in the UK', *Progress in Human Geography* 35: 483–501

Jones, R., Pykett, J. and Whitehead, M. (2011b) 'The geographies of soft paternalism: The rise of the avuncular state and changing behaviour after neoliberalism', *Geography Compass* 5: 50–62

Jones, R., Pykett, J. and Whitehead, M. (2013) *Changing Behaviours: On the Rise of the Psychological State*, Edward Elgar, Cheltenham

Kahn, M.E. (2010) *Climatopolis: How Cities will Thrive in a Hotter Future*, Basic Books, New York

Kahneman, D., Slavic, P. and Tversky, A. (eds) (1982) *Judgment Under Uncertainty: Heuristics and Biases*, Cambridge University Press, Cambridge

Kaika, M. (2005) *City of Flows*, Routledge, London

Kaplan, J.O., Krumhardt, K.M. and Zimmermann, M. (2009) 'The prehistoric and preindustrial deforestation of Europe', *Quaternary Science Review* 28: 3016–3034

Kaplan, R. (1994) 'The coming anarchy', *Atlantic Monthly*, online at: http://www.theatlantic.com/magazine/archive/1994/02/the-coming-anarchy/304670/) (accessed 3 December 2013)

Keil, R. (2005) 'Urban political ecology', *Urban Geography* 26: 640–651

Klare, M. (2002) *Resource Wars: The New Landscape of Global Conflict*, Henry Holt Books, New York

Klein, N. (2007) *The Shock Doctrine: The Rise of Disaster Capitalism*, Penguin Books, London

Klein, N. (2010) *No Logo*, Picador, New York

Knabb, R.D., Rhome, J.R. and Brown, D.P. (2005) *Tropical Cyclone Report Hurricane Katrina 23–30 August 2005*, National Hurricane Center,

20 December [Updated 10 August 2006 and 14 September 2011], available online at: http://www.nhc.noaa.gov/pdf/TCR-AL122005_Katrina.pdf) (accessed 3 December 2013)

Koné, L. (2010) 'Toxic colonialism: The human rights implications of illicit trade of toxic waste in Africa', *Consultancy Africa Intelligence*, online at: http://www.consultancyafrica.com/index.php?option=com_content&view=article&id=473:toxic-colonialism-the-human-rights-implications-of-illicit-trade-of-toxic-waste-in-africa&catid=91:rights-in-focus&Itemid=296 (accessed 6 March 2012)

Kropotkin, P. (1974) *Fields, Factories and Workshops*, Freedom Press, CA

Krueger, R. and Gibbs, D. (2008) '"Third wave" sustainability?: Smart growth and regional development in the USA', *Regional Studies* 42(9): 1263–1274

Kuhn, T.S. (1962) *The Structure of Scientific Revolutions*, University of Chicago Press, Chicago

Kunstler, J.H. (1994) *The Geography of Nowhere: The Rise and Decline of America's Man-Made Landscape*, Free Press, New York

Latour, B. [trans. Sheridan, A. and Law, J.] (1993) *The Pasteurization of France*, Harvard University Press, London

Le Billon, P. (2001) 'The political ecology of war: Natural resources and armed conflicts', *Political Geography* 20: 561–584

Le Doux, J. (1996) *The Emotional Brain: The Mysterious Underpinnings of Emotional Life*, Simon & Schuster, New York

Lerner, S. (2005) *Diamond: A Struggle for Environmental Justice in Louisiana's Chemical Corridor*, MIT Press, Cambridge

Light, A. (2001) 'The urban blind spot in environmental ethics', *Environmental Politics* 10: 7–35

Linton, J. (2010) *What is Water? The History of a Modern Abstraction*, UBC Press, Vancouver

Lomborg, B. (2001) *The Skeptical Environmentalist*, Oxford University Press, Oxford

Lorimer, J. (2012) 'Multinatural geographies for the Anthropocene', *Progress in Human Geography* 36(5): 593–612

Luccarelli, M. (1995) *Lewis Mumford and the Ecological Region: The Politics of Planning*, The Guildford Press, London

McEwan, I. (2011) *Solar*, Vintage Books, London

McNeill, J. (2000) *Something New Under the Sun: An Environmental History of the Twentieth Century*, Penguin, London

Maddison, A. (2008) 'Historical statistics for the world economy', online at: www.ggdc.net/maddison/ (accessed 28 January 2013)

Mann, M. (1984). 'The autonomous power of the state: Its origins, mechanisms and results', *European Journal of Sociology* XXV: 185–213

Marcotullio, P.J. (2007) 'Variations of urban environmental transitions: The experiences of rapidly developing Asia-Pacific cities' in Marcotullio, P.J. and McGranahan, G. (eds) *Scaling Urban Environment Challenges: From Local to Global and Back*, Earthscan, London: 45–68

Marcotullio, P.J. and McGranahan, G. (eds) (2007) *Scaling Urban Environmental Challenges: From Local to Global and Back*, Earthscan, London

Marks, K. (2008) 'The world's rubbish dump', *The Independent*, 5 February, online at: http://www.independent.co.uk/environment/green-living/the-worlds-rubbish-dump-a-tip-that-stretches-from-hawaii-to-japan-778016.html (accessed 1 July 2013)

Martinez-Alier, J. (2003) 'Urban "unsustainability" and environmental conflict' in Girard, L.F., Forte, B., Cerreta, M., De Toro, P. and Forte, F. (eds) *The Human Sustainable City: Challenges and Perspectives from the Habitat Agenda*, Aldershot: Ashgate: 89–105

Marx, K. and Engels, F. (2004) *The Communist Manifesto*, Penguin Classics, London

Massey, D. (2007) *World City*, Polity Press, Cambridge

Maxwell, S. and Frankenberger, T. (1992) *Household Food Security: Concepts, Indicators, Measurements: A Technical Review*, UNICEF and IFAD, New York and Rome

Meadows, D.H., Randers, D.L. and Behrens, W.W. III (1973) *The Limits to Growth: A Report for the Club of Rome's Project on the Predicament of Mankind*, Universe Books, New York

Merchant, C. (1990) *The Death of Nature: Women, Ecology and the Scientific Revolution*, Harper Collins, New York

Merchant, C. (2004) *Reinventing Eden: The Fate of Nature in Western Culture*, Routledge, New York

Merriam Webster (2012) 'Soil', online at: http://www.merriam-webster.com/ (accessed 2 December 2013)

Merrifield, A. (2002) *Metromarxism: A Marxist Tale of the City*, Routledge, New York

Miller, D. (2012) *Consumption and its Consequences*, Polity Press, Cambridge

Molotch, H. (1976) 'The city as a growth machine: Toward a political economy of place', *The American Journal of Sociology* 82: 309–332

Monbiot, G. (2000) *Captive State: The Corporate Takeover of Britain*, Pan Books, London

Monbiot, G. (2006) *Heat: How to Stop the Planet Burning*, Penguin Books, London

Mosley, S. (2001) *The Chimney of the World: A History of Smoke Pollution in Victorian and Edwardian Manchester*, Whitehorse Press, Harris

Mumford, L. (1961) *The City in History: Its Origins, Its Transformation, and Its Prospects*, Harvest Books, London

Nelson, M.P. (2001) 'Lynn White Jnr' in Palmer, J.A. (ed.) *Fifty Key Thinkers on the Environment*, Routledge, London: 200–205

Nerlich, B. (2010) '"Climategate": Paradoxical metaphors and political paralysis', *Environmental Values* 19: 419–442

New York Times (2012) 'The "Anthropocene" as environmental meme and/or geological epoch', *New York Times*, 17 September

Observer (2012) 'How the rise of the megacity is changing the way we live', *Observer*, 22 January: 36–37

OECD (Organization for Economic Co-operation and Development) (2010) *Cities and Climate Change: Key Messages from the OECD*, OECD, Paris

Ophuls, W. (1977) *Ecology and the Politics of Scarcity*, W. H. Freeman and Co, New York

Ophuls, W. (1997) *Requiem for Modern Politics*, Westview Press, Boulder, CO

Ophuls, W. (2011) *Plato's Revenge: Politics in the Age of Ecology*, MIT, London

Oreskes, N. (2004) 'The scientific consensus on climate change', *Science* 306: 1686

Oreskes, N. and Conway, E.M. (2010) *Merchants of Doubt*, Bloomsbury Press, London

Packard, V. (1957) *The Hidden Persuaders*, Cardinal Editions, New York

Patz, J.A., Gibbs, H.K., Foley, J.A., Rogers, J.V. and Smith, K.R. (2007) 'Climate change and global health: Quantifying a growing ethical crisis', *EcoHealth* 4: 397–405

Pearce, F. (2006) *The Last Generation: How Nature Will Take Her Revenge*, Transworld, London

Pearce, F. (2007) *When the River Runs Dry: What Happens When our Water Runs Out*, Transworld, London

Peck, J. (2010) *Constructions of Neoliberal Reason*, Oxford University Press, Oxford

Peet, R. and Watts, M. (2004) 'Liberating political ecology' in Peet, R. and Watts, M. (eds) *Liberation Ecologies: Environment, Development, Social Movements*, Routledge, London: 3–47

Peet, R., Robbins, P. and Watts, M.J. (2011) *Global Political Ecology*, Routledge, New York

Peluso N. and Watts, M.J. (eds) (2001) *Violent Environments*, Cornell University Press, Ithaca

Pepper, M., Jackson, T. and Uzzell, D. (2011) 'An examination of Christianity and socially conscious and frugal consumer behaviours', *Environment and Behaviour* 43: 274–290

Persky, J. (1995) 'The ethology of *Homo Economicus*', *Journal of Economic Perspectives* 9: 221–231

Peterson, G. (2008) 'Evaluating Ruddiman's long anthropocene hypothesis', *Resilience Science*, online at: http://rs.resalliance.org/2008/12/19/in-the-field-agu-did-the-first-farmers-stave-off-an-ice-age/ (accessed 18 May 2012)

Pichler, M. (2011) 'Sustainable palm-based agrofuels? Current strategies and problems to guarantee sustainability for agrofuels within the EU', *Policy Paper – Austrian Institute for International Affairs*, Austrian Institute for International Affairs, Zurich

Popper, K. (2002 [1950]) *The Logic of Scientific Discovery*, London, Routldge

Prudham, S.W. (2005) *Knock on Wood: Nature as Commodity in Douglas-Fir Country*, Routledge Press, New York

Prudham, S.W. (2008) 'Tall among the trees: Organizing against globalist forestry in rural British Columbia', *Journal of Rural Studies* 24(2): 182–196

Raffles, H. (2002) *In Amazonia: A Natural History*, Princeton University Press, Princeton, NJ

Richardson, K., Steffen W. and Liverman D. (eds) (2012) *Climate Change: Risks, Challenges, Decisions*, Cambridge University Press, Cambridge

Rifkin, J. (2009) *The Empathic Civilization: The Race to Global Consciousness in a World in Crisis*, Polity Press, Cambridge

Rigg, J. (2006) 'Piers Blaikie' in Simon, D. (ed.) *Fifty Key Thinkers on Development*, Routledge, London: 35–39

Robbins, P. (2007) *Lawn People: How Grasses, Weeds and Chemicals Make Us Who We Are*, Temple University Press, Philadelphia

Robbins, P. (2012) *Political Ecology*, 2nd edition, Blackwell, Oxford

Rocheleau, D. (1995) 'Gender and a feminist political ecology perspective', *IDS Institute for Development Studies* 26(1): 9–16

Rockström, J. et al (2009) 'A safe operating space for humanity', *Nature* 461(24): 472–475

Rohde, R., Muller, R.A., Jacobsen, R., Muller, E., Perlmutter, S., Rosenfeld, A., Wurtele, J., Groom, D. and Wickham, C. (2012) 'A new estimate of the average Earth surface land temperature spanning 1753 to 2011', submitted to JGR The Third Santa Fe Conference on Global and Regional Climate Change

Rohe, R.E. (1983) 'Man as geomorphic agent: Hydraulic mining in the American west', *Pacific Historian* 27: 5–16

Roy, A. (2009) 'The 21st-century metropolis: New geographies of theory', *Regional Studies* 43(6): 819–830

Royal Commission on Environmental Pollution (2007) *Twenty-Sixth Report: the Urban Environment*, London, HMSO

Ruddiman, W.F. (2001) *Earth's Climate: Past and Future*, W.H. Freeman and Co, New York

Ruddiman, W.F. (2005) *Plows, Plagues, and Petroleum*, Princeton University Press, Princeton, NJ

Sachs, W. (1999) *Planet Dialectics: Explorations in Environment and Development*, Zed Books, London

Sandberg, A. and Sandberg, T. (2010) (eds) *Climate Change – Who's Carrying the Burden? The Chilly Climates of the Global Environmental Dilemma*, Canadian Centre for Policy Alternatives, Ottawa

Sassen, S. (1991) *The Global City: New York, London, Tokyo*, Princeton University Press, Princeton, NJ

Schiermeier, Q. (2009) 'Fixing the sky', *Nature* 460: 792–795

Science Daily (2009) 'Palm oil not a healthy substitute for transfats study shows', *Science Daily*, 11 May, online at: http://www.sciencedaily.com/releases/2009/05/090502084827.htm (accessed 10 January 2013)

Scott, A. (1988) *New Industrial Spaces: Flexible Production Organization and Regional Development in North America and Western Europe*, Pion, London

Scott, J.C. (1998) *Seeing Like a State: How Certain Schemes to Improve the Human Condition Have Failed*, Yale University Press, New Haven

Shapin, S. and Schaffer, S. (1985) *The Leviathan and the Air-Pump: Hobbes, Boyle and the Experimental Life*, Princeton University Press, Princeton, NJ

Shove, E. (2003) *Comfort, Cleanliness and Convenience*, Berg, Oxford

Shove, E. (2010) 'Beyond the ABC: Climate change policy and theories of social change', *Environment and Planning A* 42: 1273–1285

Sim, S. (2006) *Empires of Belief: Why We Need More Scepticism and Doubt in the Twenty-first Century*, Edinburgh University Press, Edinburgh

Simon, D. (2007) 'Political ecology and development: Intersection, exploration and challenges arising from the work of Piers Blaikie', *Geoforum* 39: 698–707

Simon, H. (1945) *Administrative Behaviour: A Study of Decision-making Processes in Administrative Organization*, The Free Press, New York

Simon, H. (1957) *Models of Man: Social and Rational*, John Wiley & Sons, London

Simon, J.L. (1981) *The Ultimate Resource*, Princeton University Press, Princeton, NJ

Smil, V. (2004) *Enriching the Earth: Fritz Haber, Carl Bosch, and the Transformation of World Food Production*, MIT Press, Cambridge

Smith, N. (1984) *Uneven Development: Nature, Capital and the Production of Space*, Basil Blackwell, Oxford

Smith, N. (2002) 'New globalism, new urbanism: Gentrification as global urban strategy', *Antipode* 34: 427–450

Soja, E.W. (1989) *Postmodern Geographies: The Reassertion of Space in Critical Social Theory*, Verso, London

Steffen, W., Crutzen, P.J. and McNeill, J.R. (2007) 'Are humans now overwhelming the great forces of nature?', *Ambio* 36(8): 614–621

Stiglitz, J.E. (2002) *Globalization and its Discontents*, Penguin Books, London

Stoneman, D.M. (1947) *The Everglades: River of Grass*, Rinehart, New York

Striffler, S. and Moberg, M. (2003) *Banana Wars: Power, Production, and History in the Americas*, Duke University Press, Durham, NC

Swyngedouw, E. (2007) 'TechnoNatural revolutions – the scalar politics of Franco's hydrosocial dream for Spain 1939–1975', *Transactions, Institute of British Geographers* 32: 9–28

Thaler, R. and Sunstein, R. (2008) *Nudge: Improving Decisions About Health, Wealth and Happiness*, Yale University Press, London

Thomas, D.S.G. (1993) 'Sandstorm in a teacup? Understanding desertification', *Geographical Journal* 159: 318–331

Thomas, K. (1984) *Man and the Natural World: Changing Attitudes in England 1500–1800*, Penguin, London

Thoreau, D.H. (1995) *Walden: Or, Life in the Woods*, Dover Publications, New York

Thorsheim, P. (2006) *Inventing Pollution: Coal, Smoke and Culture in Britain since 1800*, Ohio University Press, Athens

Toffler, A. (1971) *Future Shock*, Pan Books, London

Twitchell, J.B. (1996) *Adcult USA: The Triumph of Advertising in American Culture*, Columbia University Press, New York

UNEP (United Nations Environment Programme)/ Habitat (2009) *Climate Change The Role of Cities: Involvement, Influence, Implementation*, UNEP/Habitat, Paris

United Nations Environmental Protection Agency (2012) *Acid Rain*, online at: http://www.epa.gov/acidrain/index.html (accessed 6 August 2012)

US Geological Survey (2011) *US Geological Survey Minerals Year Book*, online at: http://minerals.usgs.gov/minerals/pubs/commodity/myb/ (accessed 2 December 2013)

Walls, L.D. (2001) 'David Henry Thoreau, 1817–62' in Palmer, J.A. (ed.) *Fifty Key Thinkers on the Environment*, Routledge, London: 106–113

Ward, K. (1997) 'Coalitions in urban regeneration: A regime approach', *Environment and Planning A* 29: 1493–1506

Washington Post (2012) 'China's grip on world rare earth market may be slipping', 19 October

Watts, J. (2006) 'Invisible city', *The Guardian*, 15 March

Weale, A (2001) 'The politics of ecological modernisation' in Dryzek, J. and Schlosberg, D. (eds) *Debating the Earth: The Environmental Politics Reader*, Oxford University Press, Oxford: Chapter 21

Webster, P. and Burke, J. (2012) 'How the rise of the megacity is changing the way that we live', *Observer*, 22 January

Weisman, A. (2007) *The World Without Us*, Virgin Books, London

Wheen, F. (1999) *Karl Marx*, Fourth Estate, London

While, A., Jonas, A.E.G. and Gibbs, D.C. (2004) 'The environment and the entrepreneurial city: Searching for the urban "sustainability fix" in Leeds and Manchester', *International Journal of Urban and Regional Research* 28(3): 549–569

White Jnr, L. (1967) 'The historical roots of our ecological crisis', *Science* 155: 1203–1207

Whitehead, M. (2006) *Spaces of Sustainability: Geographical Perspectives on the Sustainable Society*, Routledge, Abingdon

Whitehead, M. (2008) 'Cold monsters and ecological leviathan: On the relationships between states and the environment', *Geography Compass* 2: 414–432

Whitehead, M. (2009) *State, Science and the Skies: Govermentalities of the British Atmosphere*, Wiley-Blackwell, Oxford

Whitehead, M. (2013) 'Neoliberal urban environmentalism and the adaptive city: Towards a critical urban theory and climate change', *Urban Studies* 50: 1348–67

Whitehead, M., Jones, R. and Jones, M. (2007) *The Nature of the State: Excavating the Political Ecologies of the Modern State*, Oxford University Press, Oxford

Whitehead, M., Jones, R. and Pykett, J. (2011) 'Governing irrationality, or a more-than-rational government: Reflections on the re-scientization of decision-making in British public policy', *Environment and Planning A* 43(12): 2819–2837

Wittfogel, K. (1957) *Oriental Despotism; a Comparative Study of Total Power*, Yale University Press, London

Wolch, J., Pastor Jnr, M. and Deier, P. (eds) (2004) *Up Against the Sprawl: Public Policy and the Making of Southern California*, University of Minnesota Press, Minneapolis

Wolf, E. (1972) 'Ownership and political ecology', *Anthropological Quarterly* 45(3): 201–205

World Bank (2006) *Strengthening Forest Law Enforcement and Governance: Addressing a Systemic Constraint to Sustainable Development*, Environment and Agriculture and Rural Development Departments, World Bank, Washington, DC

World Economics Forum (2013) *Global Risks 2013*, WEF, Davos

Worster, D.E. (1979) *Dust Bowl: The Southern Plains in the 1930s*, Oxford University Press, New York

Zalasiewicz, J., Williams, M., Smith, A., Barry, T.L., Coe, A.L., Brown, P.R., Brenchley, P., Cantrill, D., Gale, A., Gibbard, P., Gregory, F.J., Hounslow, M.W., Kerr, A.C., Pearson, P., Knox, R., Powell, J., Waters, C., Marshall, J., Oates, M., Rawson, P. and Stone, P. (2008) 'Are we now living in the Anthropocene?', *GSA Today* 18 February: 4–8

Index